Second Edition

SOIL FERTILITY

Henry D. Foth • Boyd G. Ellis

Department of Crop and Soil Sciences
Michigan State University
East Lansing, Michigan

CRC Press
Taylor & Francis Group
Boca Raton London New York

CRC Press is an imprint of the
Taylor & Francis Group, an **informa** business

CRC Press
Taylor & Francis Group
6000 Broken Sound Parkway NW, Suite 300
Boca Raton, FL 33487-2742

© 1997 by Taylor & Francis Group, LLC
CRC Press is an imprint of Taylor & Francis Group, an Informa business

First issued in paperback 2019

No claim to original U.S. Government works

ISBN 13:978-0-367-44839-4 (pbk)
ISBN 13:978-1-56670-243-0 (hbk)

Visit the Taylor & Francis Web site at
http://www.taylorandfrancis.com

and the CRC Press Web site at
http://www.crcpress.com

Library of Congress Cataloging-in-Publication Data

Foth, H. D.
    Soil fertility / Henry D. Foth and Boyd G. Ellis.—2nd ed.
      p. cm.
    Includes bibliographical references and index.
    ISBN 1-56670-243-7
    1. Soil fertility. 2. Fertilizers. I. Ellis, Boyd G.
    II. Title.
    S596.7.F68 1996
    631.4'2—dc20
                                               96-27155
                                             CIP

Library of Congress Card Number 96-27155

Dedication

This book is dedicated to Dr. Henry D. Foth, a teacher who always displayed a love of learning, a desire to make material understandable for students, and a dedication to bringing effective methods of teaching to the classroom. He has left a legacy of former students that he mentored who carry on in his tradition.

Preface

The second edition of *Soil Fertility* has been expanded to include more information on mineralogy as it relates to soil fertility, while keeping the original goal of the book to serve as a text for a soil fertility course at the junior/senior level and at the master's level for students who have an introductory course in soil science and several basic science courses as background. Most essential topics which are applicable to a wide variety of soil series and climatic areas are covered, but it is not intended to serve as an all-inclusive text or reference book. The treatment is an evolutionary one which considers soils as dynamic, ever-changing bodies, and relates soil fertility and management to the mineralogy of their origin.

Soils are one of the world's most important resources; protection, maintenance, and improvement of this resource is critical to maintaining a quality life on earth. It is the hope that understanding and application of information in this book will help to increase the world's food supply, but at the same time allow the soil resource to be protected for many generations.

I wish to credit Henry Foth, who had accomplished much of the revision to this edition prior to his death in 1994.

Boyd G. Ellis

Contents

CHAPTER 1

Soil Fertility and Plant Nutrition

Soil fertility and plant nutrition are two closely related subjects that emphasize the forms and availability of nutrients in soils, their movement to and their uptake by roots, and the utilization of nutrients within plants. Knowledge of soil fertility is important for the development of soil management systems that produce profitable crop yields while maintaining soil sustainability and environmental quality.

1.1 DEFINITION AND NATURE OF SOIL FERTILITY

Soil fertility is the *status of a soil with respect to its ability to supply elements essential for plant growth without a toxic concentration of any element.* Fertile soils have an adequate and balanced supply of elements sufficiently labile or available to satisfy the needs of plants.

Soil fertility can be readily altered by the application of soil amendments. Repeated application of large amounts of different nitrogen fertilizers produced the infertile and fertile soils in the foreground and background, respectively, on research plots shown in Figure 1.1. Ammonium chloride produced soil with a pH of 4.6 and a yield of only 15 bushels per acre of corn (maize) on the six-row plot in the foreground. The few stunted corn plants on the plot are magnesium deficient. The magnesium deficiency was supported by deficiency symptoms on the leaves and by chemical analysis of the leaves. The leaves also contained a toxic concentration of manganese. The development of a 4.6 pH caused an unbalanced supply of essential and toxic elements (infertility) for the corn. It was unable to effectively utilize the nutrients in the soil and fertilizer.

The plot behind the ammonium chloride-treated plot had a pH greater than 6. All plant nutrients had good availability and were balanced in this fertile soil. No element was deficient or toxic for the growth of the corn.

Figure 1.1 The sandy soil in the foreground was treated for several years with a large amount of ammonium chloride, an acid-forming fertilizer, producing a soil pH of 4.6 and an average corn yield of only 15 bushels per acre. The corn was deficient in magnesium and had a manganese toxicity.

1.1.1 Soil Fertility versus Plant Type

Because plants have evolved in different climates and on diverse soils, plants have different needs for the essential nutrients and different tolerances of the toxic elements. Cassava is native to the humid tropics and grows well on strongly acid soils high in soluble aluminum. Wheat, by contrast, originated where soils are neutral or alkaline and contain very low soluble aluminum. Many wheat cultivars (cultivated varieties) have low tolerance for soluble aluminum, and their root growth is restricted in its presence. As a consequence, an acid soil can be fertile for cassava and at the same time be infertile for wheat. Thus, when defining the characteristics of a fertile soil, it is necessary to make reference to a particular crop or class of crops. Blueberries and cassava require strongly acid soils and they might grow well on the soil with pH 4.6 in Figure 1.1.

1.1.2 The Role of Soil Fertility and Fertilizers in Soil Productivity

Soil productivity encompasses soil fertility plus all the other factors affecting plant growth, including soil management practices. Soil productivity is the *capacity of a soil to produce a certain yield of agronomic crops, or other plants, with optimum management.* All productive soils are fertile for the crops being grown, but many fertile soils are unproductive because of drought or other unsatisfactory growth factors or management practices. There is a strong positive correlation in

productive soils between fertility and other soil properties so that highly productive soils have desirable physical and biological properties, as well as high fertility.

During the early part of the 20th century, highest sugar yields in Hawaii were
obtained on the naturally fertile Vertisols and Mollisols. Sugar yields were less
on Oxisols because they were low in basic cations and deficient in phosphorus.
In recent years, sugar growers mechanized operations and greatly increased
management inputs, including use of fertilizers. After a 40-year period, Oxisols
produced the highest sugar yields owing to their excellent physical properties
(Figure 1.2). Today many intensively weathered tropical soils perform similarly
to Oxisols in Hawaii in the 1930s. Good soil management, including the use of
fertilizers, can produce crop yields on these tropical soils equal to or greater than
the yields now common on more naturally fertile soils of the temperate zones.

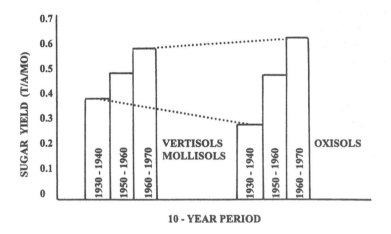

Figure 1.2 Sugar yields over a 40-year period on some Hawaiian soils. (Source: Uehara
and Gillman, 1981. Used by permission of the authors.)

1.2 HISTORICAL DEVELOPMENT

We can surmise that there has been a strong interest in the development of
plants, such as the transformation of seeds into fruit-laden plants. The American
poet, Carl Sandburg (1878-1967), said he "enjoyed watching corn grow and
wondered about how the stalk rises from the ground and eventuates into a ripening
ear with brown silk."

Interest in soil fertility likely originated with the development of agriculture.
By Roman times, many of today's soil fertility management practices were used,
including manuring, liming, crop rotations, and fallowing to build up the supply
of available nutrients. For the next 2,000 years, the accumulation of knowledge
and improvements in agricultural practice were slow.

1.2.1   Search for the Nourishment of Plants

During the latter part of the Middle Ages, the foundations of modern science were being laid, and much attention was focused on discovering how vegetation or plant growth was nourished. Many theories evolved, variously proposing water, saltpeter, soil, and the juices of the earth as the primary constituents.

Jan Baptista Van Helmont (1577-1644), a Belgian chemist, put 200 pounds of soil in an earthen vessel and grew a 169-pound willow tree in five years by adding only rainwater. The soil lost only 2 ounces, so he concluded that the tree grew because of the water.

Experimental work in Germany by chemist J. R. Glauber (1604-1668) produced the theory that saltpeter (potassium nitrate) was the principle of growth. Glauber found potassium nitrate in the earth removed from cattle barns and concluded that it came from the food or plants that the animals had eaten. He observed that the addition of potassium nitrate to soil produced large increases in plant growth. Thus, he concluded that the value of manures and other materials that increased plant growth were due to their content of saltpeter.

In the 17th century, John Woodward, an Englishman, reviewed the findings of Van Helmont and other researchers. He set up an experiment for growing spearmint with four water treatments, including rain, river water, and two kinds of sewage water. In all treatments, the plants had an abundance of water and should have grown equally, if water were the nourishment of vegetation. Plant growth, however, increased with an additional amount of terrestrial matter in the water. It was terrestrial matter — soil or earth — that nourished the plants.

Interestingly, at that time Jethro Tull (1674-1741), an Englishman, developed tillage implements to pulverize soil which produced smaller soil particles. He believed small soil particles could be taken up more readily by plant roots because root enlargement increased the pressure between roots and soil particles, thus forcing soil particles into roots.

Soon after 1800, the analytical balance was invented and it helped scientific discovery because very small quantities of chemicals and solutions could be weighed. It was discovered that plants and animals respired by the *burning of organic food* and the basic concept of photosynthesis was elucidated. A plant grown in water was found to have an ash content equal to the ash of the seed. Van Helmont's willow increased in weight largely because it fixed carbon. It was proved that plants do not ingest particles of food as animals do, but that plant roots take up nutrient elements, for the most part, as ions. An analysis of plant ash disclosed elements commonly or abundantly found in soils. Justus von Liebig (1803-1873) summarized the major findings and brought an end to the search for the nourishment of vegetation. Liebig, a German chemist, wrote that plants get most of their carbon from carbon dioxide. He said that water was a source of hydrogen and oxygen, and that the soil is a source of elements in the ash, but he erroneously believed that plants absorbed their nitrogen as ammonia from the atmosphere. Liebig's book, *Organic Chemistry in its Application to Agriculture and Physiology*, published in 1840, marks the beginning of agricultural science. Early agriculture in the U.S. was based largely on knowledge acquired in Europe.

1.2.2 Some Early Developments in America

Many of the soils along the eastern seaboard of the U.S. are naturally infertile. The soils are mostly Spodosols in New England and Ultisols farther south, as in Virginia and other states on the southeastern coastal plain. Early settlement, including that of the Pilgrims, occurred mainly on these soils that were naturally infertile for agriculture.

About that time, a native American named Squanto was kidnapped and enslaved by Europeans, and spent some time in Europe where he learned English. There, it is believed, he acquired the technique for fertilizing corn. He was returned to America and met the Pilgrims soon after their arrival.

Most of Squanto's tribe had succumbed to European diseases by the time the Pilgrims arrived and Squanto made abandoned fields available to the English puritans for growing their first crop of corn in 1621. He also taught the Pilgrims how to fertilize corn by placing a dead fish near each hill. The decaying fish produced nutrients for the corn and an abundant harvest culminated in the first Thanksgiving celebration.

For many years, the colonists produced cotton and tobacco for European markets. Worn-out land from the production of cotton and tobacco was abandoned by farmers for new land farther west in the early part of the 19th century. While the population of Virginia was declining, Edmund Ruffin, a Virginia planter, was determined to make his plantation profitable. By chance he read the book, *Elements of Agricultural Chemistry* (London, 1813) by Sir Humphrey Davy, an English chemist. Ruffin was struck by the statement, "If on wash(ing) a sterile soil it is found to contain salts of iron, or any acid matter, it may be ameliorated by the application of quick lime." Ruffin had observed plants known to contain acid, such as sheep-sorrel, growing well on soils unsuited for cultivated crops. Some areas of land along the James River were shelly, calcareous, and alkaline, and sheep-sorrel was absent. Ruffin surmised that an excess of acid was responsible for soil sterility (infertility) and that the acidity prevented the soil from being able to provide plants with the nutrients contained in manures. He began experiments with marl in 1818 and found a large increase in the yield of corn after soils were treated with marl.

In 1832, Ruffin published the results of his research in a book entitled *An Essay on Calcareous Manures*. His work led to a revival of Southern agriculture and scientists have called Ruffin the "father of soil chemistry" in America. Ruffin's book was again published in 1961, this time by Belknap Press of Harvard University as part of The John Harvard Library collection of books by important American authors.

The need for research information by farmers stimulated the development of state agricultural experiment stations. In 1862, the Department of Agriculture was established in America along with the Land Grant Agricultural Colleges by the Morrill Act. The Hatch Act in 1887 established Land Grant Experiment Stations; however, 19 states already had experiment stations. The Morrow Plots, the oldest agronomic research plots in the U.S., were established on the University of Illinois campus in 1876. The plots are shown in Figure 1.3 and contain the oldest continuous corn plot in the world. The average annual yield on the plot for the past 100 years has been about 32 bushels per acre without the application of any manure or fertilizers. This compares to about 155 bushels per acre for corn produced in a corn-oats-clover rotation and generous application of soil amendments. The plots have demonstrated that corn yields on naturally fertile soils decline rapidly without the use of manures or fertilizers and that high corn yields can be maintained with the judicious use of crop rotations, fertilizers, and lime. This famous agricultural site was declared a Registered National Historic Landmark in 1968 by the U.S. Department of Interior.

Figure 1.3 The Morrow Plots at the University of Illinois are the oldest research plots in the U.S. and contain the oldest continuous corn plot in the world. (Courtesy College of Agriculture, University of Illinois, Urbana.)

The results of these experiments and other research by educational institutions, along with the movement west to settlement on quite naturally fertile soils (mostly Alfisols and Mollisols) have produced a long period of agricultural abundance in the U.S.

1.2.3 Discovery of Essential and Toxic Elements

A chemical element is essential for plant growth when the plant is unable to complete its life cycle without it, when no other element can take its place, and when the element is directly involved in the plant's nutrition. Although nitrogen, phosphorus, potassium, and some other nutrients have been recognized as essential for a considerable length of time, the dates appear to be lost in antiquity. In

1860, iron was found essential for all plant life and many other nutrients were found essential in the 20th century (Table 1.1).

Table 1.1 Proof of Essentiality of Elements

Element	Needs proved by	Year
Iron	Sachs	1860
Manganese	McHargue	1922
Boron	Sommer and Lipman	1926
Zinc	Sommer and Lipman	1926
Copper	Sommer and others	1931
Molybdenum	Arnon and Stout	1939
Chlorine	Broyer and others	1939

Source: Adapted from Viets, 1977.

Sixteen elements have been shown to be essential for all or most higher plants. They are carbon, hydrogen, oxygen, nitrogen, phosphorus, potassium, sulfur, calcium, magnesium, iron, manganese, copper, zinc, boron, molybdenum, and chlorine. Some plants, like alfalfa, need cobalt for nitrogen fixation and some lower plants need other elements.

Iodine and selenium are essential for animals but not for plants. Plants, however, take up these elements so they are in animal feeds and forages (in some unusual cases, in toxic amounts for grazing animals).

1.2.4 Fertilizer Development

Some of the earliest materials used as sources of nitrogen, phosphorus, and potassium have included sodium nitrate from Chilean mines, bones and guano to supply phosphorus, and wood ashes and evaporite salt deposits to supply potassium. The treatment of bones with sulfuric acid to increase the solubility of the phosphorus began about 1830, and soon the modern fertilizer industry was born. In 1842, Sir John Lawes of England started the first commercial manufacture of superphosphate and, in 1850, the first mixed fertilizer was made in Baltimore, Maryland. The famous Rothamstead Experiment Station north of London, England, began field experiments in 1843 and, in 1855, the station declared that soil fertility could be maintained for many years with artificial manures or chemical fertilizers. The first commercial mining of potassium salts occurred in Germany in 1861.

The discovery in the 1880s of nitrogen fixation by bacteria of the genus *Rhizobium* allowed agriculturists to understand why legumes grow much better than other plants. Work by Fritz Haber and Karl Bosch led to the development of an efficient process for the production of synthetic ammonia in Germany in 1913. Today, N is generally the most limiting nutrient for crop production and nearly all of the N in fertilizers is derived from ammonia synthesis. The most recent fertilizer developments include more-concentrated and lower-cost materials, pesticide incorporation, and improved physical and chemical properties which have increased the efficiency of application and utilization.

1.2.5 Recent and Current Developments

After the discovery of the essential elements, research centered on their forms in soils and the factors affecting their availability for plants. Soil tests were developed and used as the basis for making fertilizer recommendations. The importance of the movement of nutrients over very short distances to roots prior to uptake was elucidated. More recently, the role of aluminum in soil acidity and as a toxic element in acid soils has received much attention. Now we know that the Ultisols on Ruffin's plantation in Virginia likely had sufficient soluble aluminum in the subsoil to inhibit root growth.

The potentially harmful consequences of applying sewage sludge to the land encouraged studies of the uptake of heavy metals by plants, their toxicity for plants, and their effects on food quality. Environmental pollution received public attention as application rates of N and P fertilizers increased. After many years of fertilization, many soils were found to have sufficient or more nutrients than are needed. Current research in soil fertility continues to stress better soil- and plant-testing methods so that fertilizer recommendations can be made for efficient crop production, prevention of elemental toxicities to plants and animals, and preclusion of environmental contamination by fertilizers.

Vescular-arbuscular mycorrhizae (ARM) are fungus roots and are present on the roots of most agricultural crops. They are being studied to develop methods for increasing the movement of nutrients from soil to and into roots, especially immobile nutrients such as phosphorus.

Current fertilizer applicators apply a uniform amount of fertilizer across fields, over-fertilizing some areas and under-fertilizing others. The result is unnecessary fertilizer costs and increased potential for environmental pollution. Site-specific fertilizer application systems are being developed. They use the Geographic Information System (GIS) to locate the fertilizer applicator in the field and the rate of application is adjusted as the applicator moves from one soil type to another. In addition, sensors are being developed that continuously measure nutrient levels as the fertilizer applicator traverses the field. The fertilizer rate is constantly adjusted according to sensor readings.

Enhanced ammonium nutrition (EAN) is a method designed to provide plants mainly with ammonium and reduce the amount of nitrate in soils. With less nitrate in soils, there is less potential for nitrate pollution of the environment.

1.2.6 Genetic Improvements to Cope with Mineral Stress

An estimated 22.5% of the world's soil has some form of mineral stress, i.e., a nutrient deficiency or element toxicity, or both. Historically, lime and fertilizers were used to remove the mineral stress in order to increase crop production. Presently, an exciting new area of research aims at modifying or improving plants so that they can better cope with mineral stress and be productive with minimal use of lime and fertilizers.

In ancient times, grape growers observed the different abilities of grape cultivars to tolerate calcareous soils in Greece, Italy, and France. As a result, grafting of desirable grape scions on efficient rootstocks became a common method of producing good wine grapes on calcareous and alkaline soils throughout the world. Today, most grapes and most fruit and nut tree tops, as well as many ornamental plants such as roses, are grafted onto efficient rootstocks. This is done to overcome mineral stress, especially micronutrient deficiencies associated with alkaline soils, and to provide protection against soil-borne root diseases. An example of the difference of soybean cultivars to grow in calcareous soil without developing iron chlorosis is shown in Figure 1.4. When the iron-efficient top is grafted onto the roots of the iron-inefficient plant (Figure 1.4), Fe-chlorosis develops. If the Fe-inefficient top is grafted onto iron-efficient roots, chlorosis does not develop.

Figure 1.4 The iron-efficient Hawkeye soybean on the left shows no iron chlorsis, but the iron-inefficient T202 cultivar on the right has severe iron chlorisis. Both are grown on the same calcareous (highly alkaline) soil. (Photograph courtesy John C. Brown.)

Scientists are now trying to identify the genes that account for tolerance of certain cultivars for toxicities and for their effective uptake of nutrients from infertile soils. Researchers are also making efforts to transfer these genes to other cultivars. Success in these efforts will have a great impact in developing countries where fertilizers and lime are expensive and supply is limited.

1.3 SOIL AS A NUTRIENT RESERVOIR

About 90% of most mineral soils other than organic soils or Histosols consist of oxygen, silicon, and aluminum. These elements are not important in plant nutrition because plants acquire their oxygen from air and water and aluminum is not an essential element. Silicon also is not considered essential even though it is sometimes beneficial for rice and sugar cane grown on soils low in silicate. The fourth most-abundant mineral soil element is iron, which plants use in very small quantities. Thus, chemically speaking, the essential nutrients that plants remove from the soil come from a relatively small percentage of the total soil.

1.3.1 Soil Nutrients versus Plant Needs

There are great differences in the amounts of the various elements in both soils and plants. For example, there are about 3,000 times more calcium and potassium than molybdenum in the soil, and plants take up about 10,000 times more calcium and potassium than molybdenum annually (Table 1.2). Of the six essential elements that plants in general take up in an amount of over 1 kg per hectare per year (multiply by 0.892 to convert to pounds per acre), the ratio of the soil's content to annual uptake ranges from a low of 50 for nitrogen to 2,000 for magnesium. For the remaining seven nutrients with uptake less than 1 kg/ha annually, the ratio of soil content to annual uptake ranges from 200 to 100,000. The nutrients absorbed in nominal amounts are usually sufficient unless some factor like soil pH causes them to be insoluble or unavailable. In general, the greater the amount that plants use, the more likely the soil supply will be insufficient for their needs.

Table 1.2 Typical Concentrations of Essential Nutrients in Mineral Soils, Annual Plant Uptake, and Ratio of Content in 10-cm Layer of Soil to Uptake

Nutrient	Soil content percent by weight	Annual plant uptake, kg/ha	Soil content, 10-cm layer, to annual uptake
Calcium	1	50	260
Potassium	1	30	430
Nitrogen	0.1	30	50
Phosphorus	0.08	7	150
Magnesium	0.6	4	2,000
Sulfur	0.05	2	320
Iron	4.0	0.5	100,000
Manganese	0.08	0.4	3,000
Zinc	0.005	0.3	2,000
Copper	0.002	0.1	1,000
Chlorine	0.01	0.06	200
Boron	0.001	0.03	400
Molybdenum	0.0003	0.003	1,000

Source: Adapted from Bohn, McNeal, and O'Connor, 1985.

Many plow layers contain 2 to 4% organic matter, which results in about 0.1 to 0.2% nitrogen. As the amount of organic matter increases, there is a corresponding increase in nitrogen content and decreasing content of those nutrients which exist primarily in the mineral fraction of the soil. About 0.9% of the world's soils are organic (Histosols). These soils, compared to mineral soils, have a high percentage of nutrients that accumulate in humus, such as nitrogen, phosphorus, and sulfur. By contrast, Histosols are commonly very deficient in potassium for crop needs. The low content of silicate minerals in Histosols may keep the amount of soluble silica in the soil solution very small, which means that the yields of sugar can increase when silicon fertilizer is applied even though, as mentioned earlier, silicon is not strictly an essential element.

1.3.2   The Soil Solution and Available Nutrients

The soil solution consists of the soil's *aqueous liquid phase and its solutes*. Nutrients exist in the soil solution mostly as hydrated ions and, to a lesser extent, as inorganic molecules and chelated compounds. Examples, respectively, include hydrated calcium ions, H_3BO_3, and iron chelate. The solutes come mainly from exchange reactions and desorption between the solid and aqueous phases, dissolution or chemical weathering of minerals, and mineralization of organic matter. The composition of the soil solution is usually determined by leaching a soil column with successive aliquots of ethanol or glycerol and determining the concentration of the nutrients in the displaced soil water.

Available nutrients consist of the *nutrient ions or compounds in forms which plants can absorb and utilize in growth*. Available nutrients are essentially the ions in the soil solution plus those ions of the solid phase that are in rapid equilibrium with the ions in solution. The available nutrients exist adjacent to the roots or in soil close enough to roots so that they can move to roots for absorption.

The amount of a nutrient in solution generally equals a small percentage of the total amount available and that amount equals a small percentage of the total amount in the soil. In the case of potassium, an acre furrow slice (2,000,000 pounds) might contain 4 pounds of available potassium in solution as hydrated ions (K^+), 400 pounds of available potassium as exchangeable ions adsorbed onto cation exchange sites, and 40,000 pounds of unavailable potassium in the crystal lattices of mica and feldspar minerals.

The relationship between the amount of exchangeable and solution potassium for a South Dakota soil is shown in Table 1.3. The soil has a high amount of

Table 1.3 Amount of Solution Potassium, Magnesium, and Calcium Relative to the Exchangeable Amount in a South Dakota Soil and Relative to the Needs of a High-Yield Corn Crop

Cation	Solution, pp2m	Exchangeable, pp2m	Solution, as percent of exchangeable	Solution, as percent of plant need
Potassium	9	711	1.2	5
Magnesium	17	1,872	0.9	42
Calcium	69	13,760	0.5	191

Source: Adapted from Black, 1968, and Barber, 1984.

available potassium, being 720 pounds per acre furrow slice. There are 711 pounds as exchangeable and 9 pounds as solution potassium, the solution potassium being equal to about 1% of the available nutrients. Assuming a high yield of corn that absorbs 174 pounds of potassium per acre (Table 1.4), an amount equal to about 5% of the plant uptake is in solution at a given instant. This situation represents a case where there is likely to be a sufficiently rapid recharge of solution potassium from the exchangeable (and unavailable) potassium to keep a corn crop well supplied throughout the growing season.

Table 1.4 Importance of Root Interception, Mass Flow, and Diffusion for the Production of 9,500 kg of Corn Grain on a Fertile Alfisol

Nutrient	Amount absorbed		Interception, kg/ha	Mass flow, kg/ha	Diffusion, kg/ha
	kg/ha	pounds/acre			
Nitrogen	190	170	2	150	38
Phosphorus	40	36	1	2	37
Potassium	195	174	4	35	156
Calcium	40	36	60	150	0
Magnesium	45	40	15	100	0
Sulfur	22	20	1	65	0

Source: Adapted from Barber, 1984.

In the case of calcium, there may be more calcium in solution than the crop needs to absorb during the full growing season. Obviously, there is little possibility for a calcium deficiency. For these three exchangeable cations, the percentage of the exchangeable nutrient that is in solution is inversely related to the energy of adsorption, being the highest for potassium and being the lowest for calcium (Table 1.3).

By contrast, nitrate (NO_3^-) is the dominant form of nitrogen in the soil solution and is taken up by plants. Nitrate is an anion and does not strongly sorb or form moderately insoluble or precipitated compounds. Essentially, all or nearly all of the usable nitrogen (nitrate) in a soil exists in solution and readily moves as the water moves. The result is that nitrate is subject to rapid movement to roots, and to uptake and rapid depletion by plants. Nitrate also is subject to rapid removal from the soil by leaching. There is a gradual recharge of nitrate into the soil solution during the growing season as unavailable N in soil organic matter is mineralized.

1.3.3 Nutrient Supplying Power

Again, using potassium as an example, the potassium ions in solution are depleted by plant uptake. Then, adsorbed or exchangeable ions desorb and enter the solution phase to maintain an equilibrium. This moderates the changes in the potassium ion concentration of the soil solution and causes a buffer effect, as potassium ions are absorbed by roots. From this example it can be seen that the nutrient supplying power includes available as well as unavailable nutrients that become available during the production of a crop. The available fraction of a nutrient is positively correlated with plant growth and soil tests attempt to extract or measure this fraction. Soil test information, along with other experimental data, is used to estimate yields without the use of fertilizer and to estimate the amount of fertilizer needed to produce a particular crop yield.

1.4 ROOT INTERCEPTION AND MOVEMENT OF NUTRIENTS TO ROOTS

Nutrients are absorbed as solutes (mainly ions) from the soil solution at root surfaces. A discussion of root interception of nutrients and the processes that move nutrients to root surfaces in position for absorption is important.

1.4.1 Root Interception

It is clear that roots occupy a very small percentage of the soil's volume and that roots are in direct contact with a small percentage of the soil's surface area. As roots elongate through soil, the roots directly encounter and intercept some of the available nutrients. The quantity of nutrients intercepted is about equal to the soil volume occupied by roots. For most field crops, roots occupy less than 1% of the 0- to 20-cm soil layer. About 1% of the available nutrients in the soil are considered to be taken up by roots because of root interception. About 4 pounds of potassium per acre would likely be intercepted by roots when the soil test for potassium is 400 pounds per acre; that is, 400 pounds of exchangeable and solution potassium in the plow layer of an acre.

1.4.2 Mass Flow

After a root becomes a resident in a soil region, solutes in the soil solution are moved to roots by mass flow caused by the convective flow of water to roots. Transpiration is the major driving force for mass flow. During mass flow, both water and solutes are moved to the root surface. The greater the concentration of a solute in the soil solution, the greater the quantity brought to the root surface in a given volume of water:

$$Q = VC$$

where Q is the quantity of solute (or ion) moved by mass flow, V is to the volume of water flow, and C is the solute concentration.

The average amount of water to produce corn in the Corn Belt is 2.5 to 3.0 million liters per hectare. Assuming that the average N concentration is 60 mg/L and 2.5 million liters of water is absorbed, 150 kg/ha of N would be moved to roots by mass flow. If the average P concentration is 0.8 mg/L, only 2 kg/ha of P would be moved to roots by mass flow. In this case, most of the N needed for the crop would be moved to roots by mass flow compared to less than 10% of the P.

When soil becomes drier as a result of transpiration and water uptake, soil hydraulic conductivity decreases, resulting in a drop in the rate of water movement and a decrease in the mass flow of nutrients to roots. If the quantity of a nutrient at the root surface is less than needed, soil drying decreases the amount of the nutrient in position for uptake and causes a reduction in uptake. When dry soil is rewetted by precipitation or irrigation, the rate of nutrient movement by mass flow increases.

If mass flow moves a greater quantity of a nutrient to the roots than is absorbed, the nutrient accumulates at the root surface. A nutrient concentration gradually develops perpendicular to roots, resulting in diffusion of the nutrient away from the roots.

1.4.3 Diffusion

Mass flow usually moves an insufficient amount of P and K to the root. Roots take up more P and K than is moved by mass flow, and produce a nutrient concentration gradient perpendicular to the root. This results in the slow diffusion of P and K toward the roots. Diffusion complements the movement of nutrients by mass flow and together mass flow and diffusion account for about 99% of nutrient movement to roots. Nutrient depletion zones adjacent to roots resulting from mass flow and diffusion are shown in Figure 1.5.

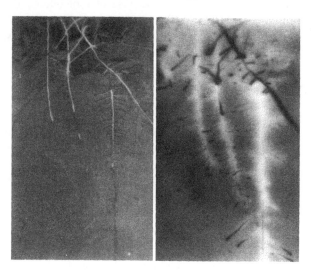

Figure 1.5 Light areas adjacent to roots depleted of radioactive rubidium. (Photograph courtesy Stanley A. Barber. Used by permission of John Wiley & Sons.)

Soil drying and low temperature reduce the rate of diffusion. The quantity of a nutrient moved by diffusion is related to (1) the cross-section or area of flow which is directly related to root length or root surface-area, A; (2) the nutrient diffusion coefficient or ease of movement through soil, k; and (3) the concentration gradient, dC/dx (change in concentration per unit change in distance). The quantity of nutrients moved by diffusion, Q, is equal to:

$$Q = Ak \, dC/dx$$

It is usually concurred that about 1% of the available nutrients are intercepted and absorbed by roots. The amount of a nutrient moved to roots by mass flow is calculated from the nutrient concentration in soil solution and quantity of water absorbed. The amount of a nutrient moved to roots by diffusion is commonly calculated by subtracting the amount of a nutrient intercepted plus the amount moved by mass flow from the total amount of nutrient taken up by the plant.

Diffusion is very important for moving nutrients to roots surfaces when their concentration in the soil solution is very low and little is moved by mass flow. It has been estimated that the distance of diffusion per day through soil at field capacity to roots is 0.13 cm for potassium ions and 0.004 cm for the phosphate ion, $H_2PO_4^-$. Considering the slow diffusion rate of nutrient ions, it takes many days for some of the ions to diffuse only 1 cm through soil. This makes it necessary for roots to be present in all soil regions from which significant nutrient uptake occurs.

1.4.4   Relative Importance of Root Interception, Mass Flow, and Diffusion

The relative importance of root interception, mass flow, and diffusion in the uptake of several nutrients by corn is shown in Table 1.4. Uptake of nitrogen due to root interception of nitrogen is 2 kg/ha, or about 1% of the 190 kg absorbed to produce a high yield of corn on a fertile soil in Indiana (Table 1.4). Available nitrogen exists mainly as soluble nitrate in solution, and the water that was absorbed moved 150 kg/ha of nitrogen to roots by mass flow. Since essentially all of the nitrate in a soil is in solution, the movement of water to root surfaces moves nearly all of the nitrate to the roots. This can result in the rapid depletion of soil nitrate by a rapidly growing and transpiring crop. The difference between 152(150 + 2) and 190 kg/ha is 38 kg/ha. This is considered the amount of nitrogen moved to roots by diffusion.

Although mass flow is very important for nitrogen uptake, diffusion is the most important for phosphorus and potassium uptake. In Table 1.4, diffusion moved about 90% of the phosphorus and 80% of the potassium to the roots. Most of the phosphorus and potassium that plants absorb during a growing season is adsorbed onto soil particles and only a relatively small percentage, as compared to nitrate, is in solution at any instant. Thus, a relatively small amount of a plant's need for phosphorus and potassium is supplied by mass flow, while the major part is supplied by diffusion. If, however, very little phosphorus and potassium were needed, that is, less than 2 kg/ha of phosphorus and less than 35 kg/ha of potassium, mass flow would have supplied more than was taken up. The relative importance of mass flow and diffusion for any given ion or nutrient depends not only on the soil nutrient characteristics but also on the amount needed by plants.

Most of the available potassium and calcium in soils exists as exchangeable cations. Calcium, compared to potassium, has much more available and corn takes up about one-fifth as much calcium as potassium. In Table 1.4, corn absorbed only 40 kg/ha; however, 60 kg were intercepted by roots and 150 kg were moved to the roots by mass flow. These calculated values are based on the assumption that there is 6,000 kg/ha of available (exchangeable) Ca and that solution concentration is 60 mg/L. If roots intercept 1% of the available calcium, 60 kg will be intercepted. If the plants absorb 2.5 million liters of water per hectare, 150 kg/ha of calcium (2,500,000 L × 60 mg/L × 1^{-6} kg/mg) will be moved to roots by mass flow. The amount from root interception and mass flow is 210 kg, which is more than five times the amount absorbed. Under these conditions, an excess of calcium is moved to the root surfaces by mass flow and a concentration gradient for calcium is established perpendicularly away from the root. This causes calcium to diffuse away from the root during part of the growing season and results in no calcium being supplied by diffusion. In alkaline soils, a high concentration of calcium and bicarbonate ions near roots can result in the precipitation and encasement of the roots with $CaCO_3$.

The situation for magnesium and sulfur is similar to that for calcium, that is, more is generally moved to roots by mass flow than is needed.

There have been efforts to classify the nutrients as mobile or immobile in terms of how well plants are generally supplied with the nutrient. Great differences in the amounts of the various nutrients that are needed complicate the situation. If little of a nutrient is required, little needs to be moved to roots and the nutrient may appear mobile in soils even though it moves with difficulty. By contrast, a nutrient may move readily but due to the large amount needed relative to the amount available, it may appear to be immobile in soils. Any nutrient that is deficient for plant growth appears to be immobile. In normal situations, however, copper, iron, manganese, phosphorus, and zinc tend to show immobile behavior. The more immobile the nutrient, the greater the importance of root length or root surface area becomes for nutrient uptake. The presence or absence of root hairs and of mycorrhiza (fungus roots) become more important for nutrient uptake as soil fertility decreases.

1.5 NUTRIENT UPTAKE FROM SOILS

Nutrient uptake from soils is essentially the product of the nutrient concentration of the soil solution and the absorbing power of roots. Root-absorbing power is affected by root length and/or surface area, kind and age of roots, plant age, temperature, plant species, and ion competition or interaction.

1.5.1   Root Morphology and Nutrient Uptake

Roots elongate through the soil by the continued division and elongation of cells in the apical meristematic zone. Cell division and elongation occur both away from and toward the shoot, which means that the root cap cells continue to form and the root stretches or gets longer. The elongation zone is made up of immature cells near the root tip where nutrients and water can freely diffuse to the center of the root and enter the transpiration stream of the xylem. The cell walls contain a lot of cellulose, which forms chain-like microfibrils. Ions and water molecules can readily diffuse into and through the intermicrofibrillar spaces. Ions absorbed into the cell walls of cortical cells can migrate centripetally by moving through the cell wall of one cell and into and through the cell wall of an adjacent cell. The cell walls are the apoplast, that part away from the plasma, and make up the apoplastic pathway. Thus, in the immature section of the root, ions and water can move across the cortex and enter the transpiration stream in the xylem vessels without crossing plasma membranes (Figure 1.6). Ion uptake in this zone is rapid and calcium uptake is restricted to this zone. Since plants are unable to translocate calcium toward the apex from older root sections, root elongation into calcium-deficient soil is inhibited.

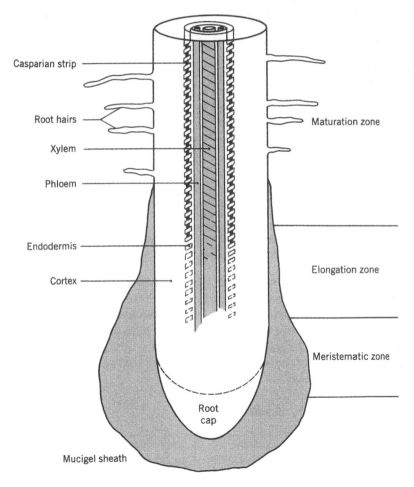

Figure 1.6 A longitudinal drawing of the apical portion of a root. In the elongation zone many cells are immature; ions and water readily diffuse to the center of the root and enter the transpiration stream in the xylem. Ion movment to the xylem in the maturation zone is inhibited by the Casparian strip in the endodermis.

Root cells quickly mature a few centimeters behind the elongation zone, and the Casparian strip forms in the endodermis that lies adjacent to the stele or central core of the root. The Casparian strip is a thickening surrounding the endodermal cells and forms a barrier at the inner cortex for the apoplastic movement of nutrients. The Casparian strip also prevents the outward flow of ions from the xylem. Along the mature section of roots, nutrients and water may move freely through the cortex via the apoplastic pathway, but they eventually encounter the endodermis, which inhibits further movement to the center of the root. Thus, to migrate across the endodermis, the nutrients must enter epidermal or cortical cells and move inward by the symplastic pathway, which consists of the cytoplasm and the interconnections between cells, called *plasmodesmata* (Figure 1.7).

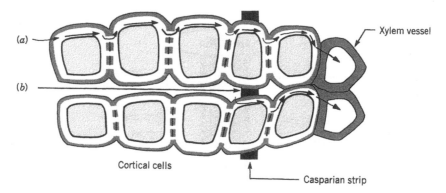

Figure 1.7 Diagram of (a) symplastic and (b) apoplastic movement of ions from the soil solution through the cortical cells. Ions that move apoplastically through the cortex must enter cortical cells and move from cell to cell symplastically (through a plasmodesma) to cross the Casparian strip in the endodermis. Plasmodesmata are cytoplasmic strands that pass through openings in some plant cell walls and provide living bridges between cells.

A mucigel sheath surrounds the root cap and extends into the elongation zone. This material, along with root exudates, forms a substrate for microorganisms in the rhizosphere. The density of microbes in the rhizosphere may be 100 times greater than that in root-free soil. These organisms have an influence on nutrient availability and nutrient uptake near the root surface. Mycorrhizal roots greatly extend the length and effective surface area of most roots and similarly influence nutrient uptake from soils.

1.5.2   Carrier Theory of Ion Uptake

Mass flow moves a large amount of calcium to root surfaces relative to plant needs. Calcium can be taken up passively along a concentration gradient between the soil solution and center of the root via the apoplastic pathway. For the other nutrients moving in the symplastic pathway, the concentration in the xylem is commonly greater than in the soil solution. Movement of ions against a concentration gradient, that is, by active uptake, requires metabolic energy. Energy appears to be needed to move ions across the plasmic membrane just inside the cell wall of epidermal or cortical cells and perhaps at other places along the symplastic route before the ions enter the transpiration stream in the xylem.

Research with large algal cells showed that cells could accumulate ions in greater concentration than in the external solution, while effectively excluding some ions in high concentration in the external solution. Plant roots also discriminate in ion uptake as shown in Table 1.5. The common crop plants accumulated much more potassium, relative to sodium, from a nutrient solution which contained 25% equivalent concentrations of each cation. Magnesium and calcium uptake were more in line with solution concentrations. The halophyte is indigenous to saline soil areas and accumulates much more sodium and much less calcium than the other plants. In general, as the concentration of an ion in the soil solution increases, uptake of that ion increases. Sometimes uptake of an ion increases with a simultaneous reduction in the uptake of some other ion. The uptake of magnesium is commonly reduced when fertilization with potassium causes an increased uptake of potassium.

Table 1.5 Discrimination of Plant Roots in Uptake of Four Cations From a Solution Containing Equivalent Concentrations of Cations

Plant	Total cations in plant, percent			
	Na	K	Mg	Ca
Buckwheat (*Fagopyrum*)	0.9	39	27	33
Sunflower (*Helianthus*)	2.3	54	17	27
Corn (*Zea mays*)	2.9	70	16	11
Potato (*Solanum*)	4.1	44	25	27
Halophyte (*Atriplex*)	19.7	39	31	10

Source: After Collander, 1941.

The carrier theory of ion absorption attempts to explain both the need for metabolic energy and ability of plants to discriminate in ion uptake. Ions readily diffuse through cell walls and encounter the plasmic membrane. At or near the plasmic membrane, ions are believed to link up with ion-specific carriers that transport the ions across the membrane and deposit them into the cell's interior. Carrier transport requires energy, and the carriers exhibit specificity to account for their discriminatory behavior.

As indicated earlier, the use of energy to accumulate ions against a concentration gradient is active uptake; most ion uptake by roots is active. In general, anions and cations are taken up simultaneously. When an excess of cations, compared to anions, is taken up, H^+ is excreted to maintain electrical neutrality in both the soil solution and the cell. With excess uptake of anions, OH^- and HCO_3^- are excreted for the same reason. Small amounts of nutrients are absorbed as molecules, including nitrogen as urea and some micronutrients as chelated compounds.

1.5.3   Nutrient Uptake and Plant Growth

The first nutrients of a germinating seed come from the seed. After germination, the roots invade the surrounding soil in search of nutrients and water. As roots leave the immediate vicinity of the seed and elongate into the surrounding soil, nutrients and water adjacent to the roots are absorbed. Water uptake causes soil drying. The drying near the root surface decreases the soil hydraulic conductivity, which results in reduced movement of both water and nutrients to the roots by mass flow. Subsequent roots will tend to grow in unoccupied soil where water and nutrients are more available. As a consequence, roots tend to space themselves rather uniformly through the soil in a lateral direction as shown in Figure 1.8. This root spacing is consistent with the generally short distance that nutrients move through soil to roots during the growing season.

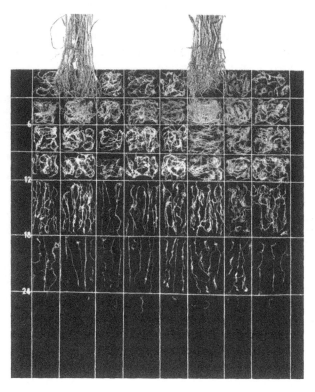

Figure 1.8 Oat roots recovered from a 10-cm thick slab of soil show a unifrom distribution laterally. The scale is in inches.

The penetration of small grain roots, such as oats and wheat, is typically less than for corn. By the end of the growing season, corn roots will usually have invaded the soil to depths of a meter or more. This is necessary to supply the water needs because of the limited mobility of water held in soil between field capacity and the wilt point. Available nutrients and roots tend to be in highest concentration in the Ap horizon, which causes a greater proportion of the nutrients to be absorbed from the Ap horizon. Also, roots may not penetrate the lower part of the root zone until late in the growing season. In general, more nutrients are absorbed from topsoils (A horizons) than the subsoils (B horizons). The reverse tends to be true for water.

Early in the development of an annual plant, root growth is relatively more rapid than shoot growth. Growth rates vary with seasons, however. By the time corn roots have 50% or their weight, only 15% of the top growth may have occurred. Relatively speaking, nitrogen, phosphorus, and potassium are taken up more rapidly in the shoots than growth occurs or dry matter accumulates. This means that delaying the applications of fertilizer, after the period of normally rapid uptake, will reduce both the opportunities for nutrient uptake and increases in yield. Of the nutrients, potassium accumulates in corn more rapidly than nitrogen, and nitrogen accumulates more rapidly than phosphorus (Figure 1.9). In addition, the amount of K in the corn shoot decreases late in the season.

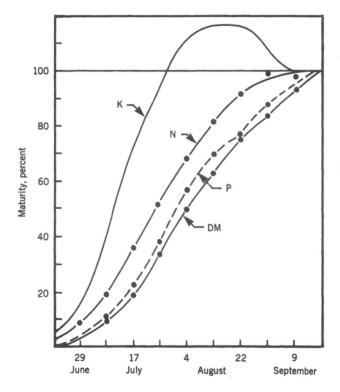

Figure 1.9 Accumulation of potassium, nitrogen, phosphorus, and dry matter of corn (maize). (Adapted from Thompson, 1957. Used by permission of author.)

Root growth in corn is especially interesting because late in the season, just before the ear develops, brace roots emerge from the lower stem nodes and branch profusely upon entering the soil. Research shows that 1 to 36% of the phosphorus in plants is absorbed by brace roots. These secondary roots can be important for nutrient uptake when the Ap horizon is occasionally rewetted by late season rains.

1.5.4 Root and Soil Interaction

It is a common observation that organisms are, in part, a product of their environment. So it is with roots. The soil environment affects the root, and the root responds with various strategies to cope with the existing soil environment. One obvious example is the excretion of oxygen by rice roots growing in flooded fields, which produces an oxygenated rhizosphere.

The differential excretion of H^+ and OH^- ions in regard to cation and anion uptake, which was mentioned earlier, modifies the pH in the immediate environment at the root surfaces. It has been observed that some plants growing in iron-stressed soils excrete additional H^+, which increases acidity and solubility of iron in the rhizosphere. Some plants excrete low-molecular-weight organic molecules that mobilize iron, which is subsequently absorbed. These mechanisms are associated with plants that are iron-efficient. A wheat cultivar that is tolerant of soil with a high content of aluminum was found to increase the pH near roots surfaces by the excretion of OH^-. This resulted in the precipitation and reduced uptake of aluminum. Thus, the interactions of roots with soils are very complex, and roots are active, not passive, in modifying their environment. This helps them to better cope with nutrient deficiencies and element toxicities.

1.6 IMPORTANCE OF FERTILIZERS

Soils can sustain a low level of crop yields for centuries when soil management inputs are minimal. This level is in the range of 8 to 10 bushels/acre or about 1,000 to 1,200 kg/ha of grain and is comparable to wheat yields in Britain between the years 1100 and 1350. Note in Figure 1.10 that between 1350 and 1550 wheat yields increased slowly as grass was used to improve privately-owned lands that had become exhausted during the feudal period. Over the next 350 years, yields increased slowly through trial and error. The sharp rise in yields since 1900 reflects the application of scientific knowledge about pest control, improved varieties, and soil management.

1.6.1 Role of Fertilizers in Food Production

Large increases in crop yields have occurred in many countries since World War II. Corn yields nearly doubled from 1939 and 1961 in the Corn Belt of the U.S. Partitioning of the factors that caused the yield increase showed that improved varieties accounted for 36% of the increase, followed by the use of fertilizer for 31%. Several studies have concluded that fertilizer use accounts for 25 to 50% of the grain production in the U.S.

Since the industrial revolution began more than 200 years ago, the accumulation of wealth has produced investments that have increased soil fertility in

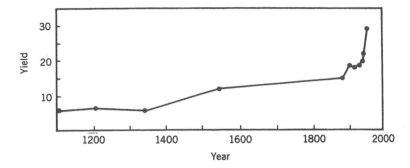

Figure 1.10 Wheat yields in Britain from 1100 to 1960 in hundred weights per acre. (Adapted from Jacks, 1962.)

many nations. A study by the Food and Agriculture Organization of the United Nations (FAO) of 41 countries found a positive relation between the amount of fertilizer used and the value index (yield) of crops produced per hectare. The most developed countries tended to use fertilizer at the highest rates. Thus, it is seen that the fertility of soils and the use of fertilizers play an important role in world food production and are related to the overall economic activity in a country.

1.6.2 Future Needs

In general, crop yields in the developed world are high and agricultural soils have high fertility. Most of this fertility is the result of intensive use of fertilizers. Also, much of the agricultural land of developed countries is in the temperate zone and these soils are quite fertile naturally (examples of such soils are Alfisols, Mollisols, and Vertisols). Frequently, the cost of fertilizer is a small part of the total production costs so that more fertilizer is applied than is needed just to make sure that yields, and net profit, will not be limited by a lack of soil fertility.

The substitution of chemical fertilizers for animal manures on farms without livestock has left less land for disposal of the large amount of manures and sewage sludge which are available for land application. The result has been high rates of application and an overload of the environment with nutrients, especially nitrogen and phosphorus. A major problem today is the overuse of animal manures and fertilizers and a resultant deteriorating environment quality.

Crop yields and soil fertility are generally much lower in the developing countries and there is a need for increased use of fertilizer to increase yields and produce sufficient food for expanding populations. Part of the solution is economic, that is, making it profitable for farmers to use more fertilizer and produce more food. Another answer is the development of superior cultivars that can yield well under mineral stress and other unfavorable conditions.

In developed and developing nations, there is a need for more efficient fertilizer use to reduce food production costs and reduce environmental pollution. The soil fertility principles expressed in this book should be useful to all persons striving to increase the availability of food in the world, and at the same time, striving to maintain soil sustainability and to maintain or improve environmental quality.

REFERENCES

Auer, L., E. O. Heady and F. Conklin. 1966. Influence of Crop Technology on Yields. *Iowa Farm Science.* 20:13–16.

Barber, S. A. 1984. *Soil Nutrient Bioavailability.* John Wiley & Sons, New York.

Black, C. A. 1968. *Soil-Plant Relationships.* John Wiley & Sons, New York.

Bohn, H. L., B. L. McNeal and G. A. O'Connor. 1985. *Soil Chemistry.* John Wiley & Sons, New York.

Brown, J. C., R. S. Holmes and L. O. Tiffin. 1957. Iron Chlorosis in Soybeans as Related to the Genotype of Rootstalk. *Soil Sci.* 86:75–82.

Ceci, L. 1975. Fish Fertilizer: A Native North American Practice? *Science.* 188:26–30.

Collander, R. 1941. Selective Absorption of Cations by Higher Plants. *Plant Physiol.* 16:691–720.

Ellis, B. G., B. D. Knezek and L. W. Jacobs. 1983. The Movement of Micronutrients in Soils. In *Chemical Mobility and Reactivity in Soil Systems.* Soil Sci. Soc. Am., Madison, WI.

Foth, H. D. 1962. Root and Top Growth of Corn. *Agron. J.* 54:49–52.

Foth, H. D. 1990. *Fundamentals of Soil Science.* 8th ed. John Wiley & Sons, New York.

Harmsen, K. 1991. Soil Fertility Problems in the Developed and Developing World. In *Interactions at the Soil Colloid-Soil; Solution Interface,* pp. 493–506. G. H. Bolt, et al., Eds. pp. 493-506. Kluwer Academic Publishers, Boston.

International Fertilizer Development Center and United Nations Industrial Development Organization. 1979. *Fertilizer Manual.* Ref. Manual IFDC-R-1. Muscle Shoals, AL.

Jacks, G. V. 1962. Man: the Fertility Maker. *J. Soil Water Con. J.* 17:146, 148, 176.

Mengel, K. and E. A. Kirkby. 1982. *Principles of Plant Nutrition.* 3rd ed. Int. Potash Institute, Bern, Switzerland.

Robertson, J. A., B. T. Kang, F. Ramirez-Paz, C. H. E. Werkhover and A. J. Ohlrogge. 1966. Principles of Nutrient Uptake From Fertilizer Bands VII: Uptake by Brace Roots of Maize and Distribution Within Leaves. *Agron. J.* 58:293–296.

Ruffin, E. 1832. *An Essay on Calcareous Manures.* J.W. Campbell, Petersburg, Virginia. (Reprinted by the Belknap Press of Harvard University as part of the John Harvard Library Book collection in 1961 with J. C. Sitterson as editor.)

Sayre, J. D. 1948. Mineral Accumulation in Corn. *Plant Physiol.* 23:267–281.

The White House. 1967. *The World Food Problem.* U.S. Goverment Printing Office, Washington, D.C.

Thompson, L. M. 1957. *Soils and Soil Fertility.* 2nd ed. McGraw-Hill, New York.

Uehara, G. and G. Gillman. 1981. *The Mineralogy, Chemistry, and Physics of Tropical Soils with Variable Charge Clays.* Westview Press, Inc. Boulder, CO.

Viets, F. G. 1977. A Perspective on Two Centuries of Progress in Soil Fertility and Plant Nutrition. *Soil Sci. Soc. Am. J.* 41:242–249.

Wild, A. 1988. *Russell's Soil Conditions and Plant Growth.* 11th ed. Longmans, London.

Wolcott, A. R., H. D. Foth, J. F. Davis and J. C. Shickluna. 1965. Nitrogen Carriers: Soil Effects. *Soil Sci. Soc. Am. Proc.* 29:405–510.

Wright, M. J. Ed. *Plant Adaptation to Mineral Stress in Problem Soils.* Proc. workshop held at Beltsville, MD in 1976. USAID, Washington, D.C.

CHAPTER 2

Charge Properties

The discovery of negative charge in soils occurred about 1850 in England by a Yorkshire farmer, Harry Stephen Thompson. He reported that leaching a soil with an ammonium sulfate solution produced a leachate that contained calcium sulfate. The ammonium ions in the solution exchanged for calcium ions in the soil. The cation exchange reaction that occurred is represented as:

$$(NH_4)_2 SO_4 + CaX = (NH_4)_2 X + CaSO_4 \qquad (2.1)$$

where X is the exchanger surface. Adsorbed calcium ions on soil particles were exchanged by ammonium ions from the ammonium sulfate solution, and leached from the soil with sulfate ions. This showed that the soil preferentially adsorbed cations as compared to anions, and that the soil had *negative charge.*

Within the past fifty years considerable knowledge has been accumulated about positive charge and anion exchange. In most soils, the positive charge is much less than the negative charge. As soils become extremely weathered, however, the positive charge becomes more nearly equal to the negative charge. In a few soil horizons, the positive charge exceeds the negative charge and the net charge is positive. Charged sites adsorb essential anions and cations that are exchangeable and available to plants making the type and amount of charge importantly related to soil fertility and plant growth.

2.1 CONSTANT CHARGE

Constant charge results from structural imperfections, such as isomorphous substitution in mineral structures. It is a permanent characteristic of some minerals and is not affected by the external environment. Isomorphous substitution of cations is common in phyllosilicate minerals that have layered structures composed of shared octahedral and tetrahedral sheets. Substitution is especially common in the 2:1 layer structures.

2.1.1   The 2:1 Layer Structure

The 2:1 layer consists of one octahedral sheet sandwiched between two tetrahedral sheets. The tetrahedral sheet is composed of silicon-oxygen tetrahedra, SiO_4^{4-}, that form six-membered rings by the sharing of basal oxygen ions. The apical oxygen ions have one unsatisfied negative bond as shown in Figure 2.1. The tetrahedral sheet has the composition $Si_2O_5^{2-}$.

Figure 2.1 Model of silicon-oxygen tetrahedral sheet showing six-membered rings of tetrahedra that share basal oxygen ions and apical oxygen ions with one unsatisfied negative bond.

Two tetrahedral sheets, with their apical oxygen facing inward, form the skeleton for the 2:1 layer, see Figure 2.2. The apical oxygen of the tetrahedral sheets are shared with octahedral plane cations which neutralize their charge. The remaining charge of the octahedral cations is neutralized by OH^-. Each octahedral plane cation is in six-coordination by four O and two OH ions. Silicon is the tetrahedral plane cation and aluminum is the octahedral plane cation in the 2:1 layer structure shown in Figure 2.2.

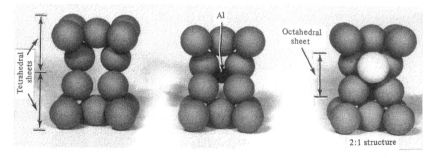

Figure 2.2 Models illustrating the 2:1 layer structure consisting of an octahedral sheet sandwiched between two tetrahedral sheets. On the left, two silicon-oxygen "sheets" facing each other. The center model shows the location of the octahedral plane cations. The octahedral Al^{3+} is surrounded by four apical oxygen ions and two hydroxyl (the OH in the rear is hidden) to place the cation in six-coordination in an octahedron. The Al^{3+} shares half a valence bond from each of the six surrounding anions.

The 2:1 structure extends outward in all directions to form plate-shaped layers whose upper and lower planar surfaces consist of oxygen that is tetrahedrally coordinated with silicon. No isomorphous substitution is shown and the internal structure is neutral. The vertical bonding within the 2:1 layer structure is O–Si–O–Al–O–Si–O and is readily apparent from observation of the model in Figure 2.3. A few hydroxyls occur within the octahedral layer and some are exposed along the layer edges of clay particles in soils.

Figure 2.3 Model of 2:1 layer structure of phyllosilicates. The upper and lower planar surfaces are oxygen. The model has silicon as the tetrahedral plane cation and aluminum as the octahedral plane cation. Only two of the three possible octahedral interstices are occupied by aluminum, making the mineral dioctahedral.

A single 2:1 layer is about 1.0 nm thick. A 2:1 phyllosilicate clay particle is visualized as consisting of a stack of many plate-shaped layers that extend outward in all directions, similar to sheets of stacked plywood. Unweathered mica particles exhibit a platy or layer structure on a macroscale, which is the case for phyllosilicate clay particles on a microscale.

2.1.2 Isomorphous Substitution

The internal structure of a 2:1 layer is neutral when the tetrahedral cations are all Si^{4+} and the octahedral cations are all Al^{3+}, as shown in Figure 2.3. This mineral is pyrophyllite and is unimportant in soils. Pyrophyllite represents an end member of a series of 2:1 layer phyllosilicates with constant charge resulting from isomorphous substitution.

Isomorphous substitution of similar-sized atoms occurs during mineral formation without changing the mineral structure. Isomorphous substitution of tetrahedral and octahedral plane cations by cations of lower valence creates an unsatisfied negative charge that is a permanent feature of the mineral. Substituted ions are generally no more than 15% different in dimension and differ by no more than one valence. Aluminum (Al^{3+}) is in size between Si^{4+} and Mg^{2+} and Fe^{2+} and is capable of isomorphous substitution with any of these. When atoms of different valences substitute for each other, electrical neutrality is maintained by substitution elsewhere within the mineral structure or by the retention of adsorbed ions on external surfaces.

Important isomorphous substitutions in soil minerals include the substitution of octahedral Al^{3+} by Mg^{2+} in montmorillonite and by Fe^{2+} in nontronite. During the formation of micaceous minerals, Al^{3+} substitutes for tetrahedral Si^{4+} in the tetrahedral cation plane. This charge originates closer to the planar surface than the charge originating in the octahedral cation plane. Adjacent layers with negative charge originating in the tetrahedral cation plane, compared to the charge originating in the octahedral cation plane, attract each other with greater energy and attract cations in solution with more energy.

Substitution in crystal lattices of an ion with higher valence, such as Ti^{4+} for Fe^{3+}, produces constant positive charge. This has been reported for iron oxide minerals. The amount of charge is small and is considered an insignificant charge source in soils.

Constant charge is a permanent characteristic of certain minerals, and being unaffected by external factors, such as pH, it is also called permanent charge. Most of the charge of minerals in temperate region soils is constant or permanent negative charge from 2:1 layer phyllosilicates.

2.2 VARIABLE CHARGE

Variable charge originates from preferential adsorption of certain ions on particle surfaces. The charge is both negative and positive and is pH dependent. Depending on pH, the net variable charge of mineral particles is negative, positive, or zero.

2.2.1 Variable Charge Minerals

Iron and aluminum minerals, such as hematite [$Fe(OH)_3$], goethite [FeOOH], and gibbsite [$Al(OH)_3$] have hydroxylated surfaces. Variable positive and negative charge originate by protonation and deprotonation, respectively, of surface exposed hydroxyls. Surface exposed OH, –Fe–OH for example, when encountering a solution with high H^+ activity, protonate producing $–Fe–OH_2^+$, a site of positive charge.

$$-Fe-OH + H^+ = FeOH_2^+ \qquad (2.2)$$

Deprotonation at high pH due to high hydroxyl activity results in development of negative charge.

$$-Fe-OH + OH^- = -Fe-O^- + H_2O \qquad (2.3)$$

Hematite and gibbsite are minerals with the cations in octahedral coordination surrounded by anions, OH, bonded to two cations. Along the edge, however, some of the OH⁻ are bonded to only one cation and these hydroxyl play the major role in the development of variable charge (see Figure 2.4).

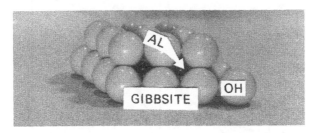

Figure 2.4 Model of gibbsite. Most of the Al are in six-coordination and bonded to two Al, except along the edges. (From *Fundamentals of Soil Science*, 7th ed., H. D. Foth. © 1984. Reprinted by permission of John Wiley & Sons.)

A small amount of variable charge develops along the edges of phyllosilicates where OH⁻, singly coordinated with aluminum, are exposed:

$$\text{clay edge } -\text{Al}-\text{OH} + \text{H}^+ = \text{clay edge } -\text{AlOH}_2^+ \tag{2.4}$$

$$\text{clay edge } -\text{Al}-\text{OH} + \text{OH}^- = \text{clay edge } -\text{Al}-\text{O}^- + \text{H}_2\text{O} \tag{2.5}$$

Protonation and deprotonation are the major or sole sources of charge in 1:1 layer phyllosilicates, including kaolinite and halloysite, and iron and aluminum oxide minerals. These minerals have a small amount of negative and positive charge that dominates their charge properties and are called variable charge minerals. Variable charge is a relatively small source of charge in 2:1 layer minerals because of the relatively large amount of charge from isomorphous substitution and the very limited number of surface Al–OH for protonation and deprotonation.

2.2.2 Zero Point of Charge

The nature of adsorption determines the variable charge, and the adsorbed ions, H⁺ and OH⁻, are called the *charge determining ions*. Because the net charge is related to pH, which governs protonation and deprotonation, the charge has also been called pH dependent. At a particular pH, the negative charge and positive charge are equal and the net charge is zero. Then, the mineral is at the zero point of charge, the ZPC. The pH of the ZPC, pH_0, of allophane is near pH 6 as shown in Figure 2.5.

The 2:1 layer phyllosilicates in soils also develop variable charge from edge-exposed –Al–OH, giving them both constant and variable charge. Such minerals, however, are called constant charge minerals because the constant charge is many times greater than the variable charge and the constant negative charge dominates

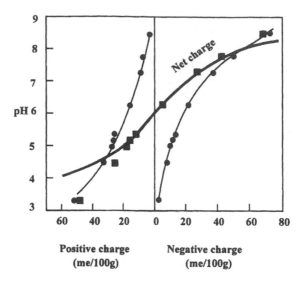

Figure 2.5 Effect of pH on the net charge of allophane weathered from pumic. (Modified from Wada, 1989. Used by permission of Soil Science Society of America.)

the mineral's properties. They remain highly negatively charged and do not develop a zero net charge within the pH range found in most soils.

2.2.3 Variable Charge of Soil Organic Matter

Negative charge originates from the deprotonation of OH of several active groups, such as carboxyl ($-COOH$) and phenolic ($-C_6H_4OH$). These groups account for about 85% or more of the negative charge of soil organic matter (SOM). Phenolic groups are weaker acids than carboxyl groups and contribute charge at higher pH, as compared to carboxyl. The COOH groups made up 76% of the cation exchange capacity (CEC) in five New York soils. In the pH range of most soils (pH less than 8.0) carboxyl groups provide most of the negative charge of SOM that is strongly pH dependent. Deprotonation is enhanced by increased OH⁻ activity producing pH-dependent charge as follows:

$$-COOH + OH^- = -COO^- + H_2O \qquad (2.6)$$

The CEC increases several times from pH 3 to 10.

Little, if any, protonation of active groups and positive-charge formation occurs for humus. Soil organic matter has high variable negative charge and, within the pH range of soils, remains strongly net negatively charged.

2.3 DETERMINATION OF CHARGE

The negative charge of materials and soils is neutralized by adsorbed cations and the positive charge is neutralized by adsorbed anions. These weakly or nonspecifically adsorbed ions are exchangeable. The amounts of exchangeable cations and anions adsorbed from solution are equal to the cation exchange and anion exchange capacities, respectively.

2.3.1 Cation Exchange Capacity

The cation exchange capacity (CEC) is *the sum of exchangeable cations that a soil or other material can adsorb at a specific pH.*

Isomorphous substitution produces constant negative charge for cation adsorption giving rise to constant cation exchange capacity (CEC_c). The negative charge resulting from deprotonation gives rise to variable cation exchange capacity (CEC_v). The sum of the CEC_c and the CEC_v equals the total CEC, CEC_t.

The CEC is expressed as cmol(+)kg^{-1}. Each centimole per kilogram represents adsorbed exchangeable cations whose sum of positive charge is equal to 6.02×10^{21} charges. The major nonspecific adsorbed exchangeable cations include Ca, Mg, K, Na, and Al.

Many older publications express the CEC as milliequivalents per 100 grams of oven-dry soil. The numerical amount of CEC is the same for both systems of expression.

Because part of the negative charge of soils is variable or pH dependent, the CEC must be determined at a standard or specified pH to make valid comparisons between soils. The CEC is commonly determined at pH 7.0 by treating the soil with a normal ammonium acetate solution adjusted to pH 7.0. Ammonium ions (NH_4^+) replace the adsorbed or exchanged cations by mass action. The excess ammonium, along with the exchanged or displaced cations, are leached out with alcohol adjusted to pH 7.0. The amount of ammonium retained by the soil is measured and equated to the CEC.

A second procedure determines CEC at pH 8.2 using NaOAc and BaCl$_2$ plus triethanolamine (TEA) solution. The NaOAc replaces the Ca, Mg, K, and Na, which are measured and added to the amount of acidity neutralized by the TEA.

A third procedure measures the CEC at the soil's current pH by leaching the soil with an extracting solution of unbuffered KCl or CaCl$_2$ to replace the exchangeable cations and leach them from the soil. Then, the amount of each exchangeable cation in the leachate is determined and the sum of the cations is equated to the CEC. This method is important for determining the CEC of soils at their present or natural pH, particularly for acid soils.

2.3.2 Anion Exchange Capacity

The soil's positive-charge capacity is measured by determining the amount of negative charge on the nonspecific adsorbed anions. Major nonspecific adsorbed anions include nitrate and chlorine.

The amount of positive charge is expressed as the anion exchange capacity (AEC). The AEC is *the sum of exchangeable anions that a soil, or other material, can adsorb at a specific pH*. The AEC is expressed as cmol(–)kg^{-1}. The AEC and the CEC are commonly expressed as centimole of charge per kilogram, cmol$_c$ kg^{-1}.

2.4 THE CHARGE OF SOIL ORGANIC MATTER

The organic matter of soils can be divided into three fractions: (1) living organisms (flora and fauna) or biomass, (2) the remains of dead plants and animals, and (3) humified organic matter or humus. The living and dead remains (nonhumus fraction) are the major energy source for the microbial population that converts the organic matter added to soils into humus.

2.4.1   Nature of Humus

Most SOM is humus composed of polymers with high molecular weight. Humus has three major components: humin, humic acid, and fulvic acid. The humus fraction that is not solubilized by treatment with NaOH is humin. The NaOH soluble portion contains the humic and fulvic acid fractions. The NaOH soluble fraction, precipitated by treatment with concentrated HCl and brought to a pH of 1, is humic acid. The remaining NaOH soluble fraction is fluvic acid.

The maximum charge of humic acid from the total carboxyl groups is about 360 cmol$_c$ kg^{-1} and of the phenolic groups is 310 for a total maximum possible charge of 670 cmol$_c$ kg^{-1}. Comparable values for fulvic acid are 820, 300, and 1120 cmol$_c$ kg^{-1}, respectively. These values are averages for many soils throughout the world and on a mass basis are much higher than mineral soil fractions.

2.4.2   Effect of Decomposition on Negative Charge

The nonhumus fraction comprises only several percent of the total SOM in soils and has a CEC of the order of 15 to 35 cmol$_c$ kg^{-1}. As a result, the nonhumus fraction contributes only a minor amount to the total negative charge of SOM. As organic matter decomposes in the soil, the nonhumic fraction is mineralized and there is an increase in acidic groups resulting in an increase in charge. Plant residues incubated and decomposed 452 days increased the carboxyl content of decomposing plant residues from 28 to 134 cmol kg^{-1} and the CEC from 25 to 82 cmol$_c$ kg^{-1}.

The bulk of humus in mineral soils is generally well decomposed and has a mean residence time from hundreds to thousands of years. Such humus has quite high variable charge and CEC, averaging about 200 cmol$_c$ kg^{-1} and ranging from 100 to 400 cmol$_c$ kg^{-1} at pH 7. In many A horizons and plow layers the organic matter contributes about half or more to the CEC$_t$ and to a lesser proportion of the CEC$_t$ of subsoils, due to the generally decreasing amount of organic matter with increasing soil depth.

2.5 THE MINERAL WEATHERING SEQUENCE AND CHARGE DEVELOPMENT

The charge of soil minerals comes mainly from the clay fraction. Silt makes a significant contribution in some silty soils, while little, if any, contribution comes from the sand fraction. The clay fraction of soils is usually dominated by fine-sized secondary minerals, *clay minerals*, with high specific surface and sites of both negative and positive charge.

The clay fraction typically contains several clay minerals referred to collectively as the *clay-mineral suite*. The one or two minerals in greatest abundance tend to dominate the charge properties of the mineral fraction.

The consideration of the origin of a clay mineral suite and the source of the charge will focus on the generally accepted weathering sequence of mica to vermiculite and smectite to hydroxy-Al interlayered vermiculite and smectite to kaolinite and, finally, to gibbsite. During this weathering sequence, clay-sized 2:1 layer minerals with high constant negative charge and low variable negative and positive charge, tend to form in minimally and moderately weathered soils. The net charge is predominately negative and the CEC_t is high. With time, if weathering conditions are favorable, these clay minerals are gradually transformed into 1:1 layer and oxide minerals with low variable negative and positive charge, characteristic of clay minerals in intensively weathered soils. Depending on pH, the net charge of these minerals can be negative or positive.

The weathering sequence approach provides a unifying concept for understanding the origin of the charge characteristics of soils. The mineral groups to be considered include 2:1 and 1:1 layer phyllosilicates, aluminosilicate minerals with short-range order, and oxide minerals.

2.6 CONSTANT CHARGE 2:1 LAYER PHYLLOSILICATES

The 2:1 layer phyllosilicates have high constant negative charge and low variable negative and positive charge. Properties are dominated by the constant charge and they are called constant charge minerals. They are commonly the dominant clay-minerals in minimally and/or moderately weathered soils. Many of these soils are Entisols, Inceptisols, Alfisols, Mollisols, and Vertisols.

2.6.1 Mica, Weathered Mica, and Vermiculite

The most common soil micas, muscovite and biotite, are generally inherited from the parent material and are ubiquitous in soils. Sedimentary rocks, such as shale and slate, and igneous and metamorphic rocks, such as schist, gneiss, and granite, are high in mica content. Micaceous minerals are abundant in the till from which many soils of the north-central U.S. developed. Parent materials derived directly from these rocks or sediments contain large amounts of mica.

Micas are 2:1 layer phyllosilicates, with tightly held nonhydrated interlayer cations that balance a high layer or lattice charge. The micas, biotite and muscovite, have about one-fourth of the tetrahedral Si^{+4} substituted by Al^{+3}, resulting in high constant negative charge in the tetrahedral sheet. The Al^{3+} is slightly larger than Si^{4+}, the ions having 0.051 and 0.042 nm ionic radii in the mineral, respectively. The substitution causes a slight distortion in the structure.

Each substitution creates a site for the retention of a cation. The tetrahedral sheet charge in mica is balanced by nonhydrated K ions in the interlayer space. The K^+ occupy cavities on the outer plane of tetrahedral sheets coordinated with twelve oxygen (six oxygen from the tetrahedral sheets of two adjacent layers). The common attraction of the interlayer K^+ for two adjacent negatively charged tetrahedral sheets results in the layers being held together tightly and the layers are nonexpanding. A model of muscovite is shown in Figure 2.6a.

In muscovite, the octahedral sheet cation is Al. It is surrounded by oxygen and hydroxyl, and the sheet is neutral. The dioctahedral nature of the model in Figure 2.6a can be observed by noting that two of every three octahedral cation locations or interstices are occupied. The substitution of Al in the tetrahedral sheet [Si_3Al], and no substitution in the octahedral sheet [Al_2], is reflected in the idealized formula for muscovite:

$$K[Si_3Al][Al_2]O_{10}(OH)_2$$

Biotite has Mg^{2+} and Fe^{2+} as the dominant cations in the octahedral sheet. The divalent cations occupy all of the interstices of the octahedral sheet, making the mineral trioctahedral. This is reflected in the idealized formula of biotite.

$$K[Si_3Al][Mg, Fe]_3O_{10}(OH)_2$$

Both biotite and muscovite have similar layer charge, resulting from isomorphous substitution in the tetrahedral sheets, and are nonexpanding. Oxidation of Fe^{2+} in the octahedral sheet during weathering produces structure instability and causes biotite to weather much more rapidly than muscovite. Muscovite/biotite ratios in the sand fraction of soils tend to increase with time, and biotite is commonly absent in only moderately weathered soils. Muscovite, by contrast, persists in some intensively weathered soils.

During the weathering of mica, interlayer K is lost first along the particle edges. The loss of interlayer K^+ results in a loss of positive charge for holding adjacent negatively charged 2:1 layers together and creates sites where unsatisfied negative charge attracts exchangeable cations. This creates CEC_c and wedge-shaped voids or spaces into which hydrated cations in solution are attracted and adsorbed. The loss of interlayer K parallels the increase in CEC_c. A model of partially weathered mica, Figure 2.6b, shows two exchangeable cations adsorbed on an external planar surface and one exchangeable cation adsorbed on the edge.

The hydrated exchangeable cations shown in Figure 2.6b represent cations surrounded by six water molecules. This hydrated cation model is representative of the relative size of unhydrated Ca^{2+} and Mg^{2+}, which are smaller than K^+ in the nonhydrated state. Hydrated Ca and Mg ions however, are larger than unhydrated K ions. The interlayer space is too small for the migration of exchangeable hydrated cations into it.

Protons, abundant in acid soils, are effective in replacing interlayer K. Eventually, there is complete removal of interlayer K and maximum development of negative charge and CEC_c in the tetrahedral sheets. The complete loss of interlayer K is associated with the separation of layers and expansion of the interlayer space; the weathered mica has been transformed into vermiculite. The expanded interlayer space is about 0.4 nm thick.

The high charge in the tetrahedral sheets adjacent to the interlayer space attracts hydrated Ca and Mg ions and other cations, which move into and out of the expanded interlayer space. Specimen vermiculites have shown high selectivity of Mg over Ca and selectivity of Ca over Na. Soil vermiculites have shown less selectivity for hydrated cations in the interlayer space. Calcium and Mg have high hydration energies and are pulled into the interlayer space with their accompanying hydration water. A model of vermiculite is shown in Figure 2.6c.

The loss of interlayer K is not uniform between all mica layers and particles are formed containing interlayer spaces that have lost all of their K and interlayer spaces where only part of the K has been weathered out. This produces an interstratified clay with a random or orderly stacking of layers of different kinds. This results in a range in the charge properties of different mineral samples and soil clays.

The CEC_t of vermiculite is about the highest found in soil clay minerals with the CEC_c greatly exceeding the CEC_v. The idealized calculated CEC of dioctahedral vermiculite altered from muscovite is about 250 $cmol_c$ kg^{-1}. Cation exchange capacities of soil vermiculites are commonly given as 120 to 150 $cmol_c$ kg^{-1}. Soils clays are neither ideal nor pure and many other CEC_t values for vermiculite have been reported, including 124 $cmol_c$ kg^{-1} with about 90% being CEC_c and about 10% being CEC_v.

Charge distribution for several mineral specimens is given in Table 2.1. The CEC_t was determined at pH 8.2 with barium as the index cation. CEC_c was determined by displacement of cations with unbuffered KCl. The difference between the CEC_t and CEC_c is the CEC_v. The AEC was determined by phosphate adsorption. The data in Table 2.1 show much greater CEC_t values for vermiculite than for weathered mica, and in both cases the CEC_c greatly exceeds the CEC_v. Anion exchange capacity of both mica and vermiculite is low. The data in Table 2.1 gives the AEC of vermiculite as zero; however, there is normally a small amount. The mineral specimens are not pure; each sample reported in Table 2.1 contained over 50% of the mineral specified.

Vermiculite and weathered mica can be converted to mica by treatment with a solution high in K$^+$ in the laboratory. The treatment results in replacement of interlayer K, loss of CEC_c, and collapse of the layers.

When a high concentration of hydrated K$^+$ occurs near CEC_c sites, caused by the application of K fertilizer, K$^+$ reenter the wedge-shaped interlayer zones at particle edges and become fixed and unavailable. This fixation is related to the strong attraction of adjacent negatively charged tetrahedral sheets of the layers

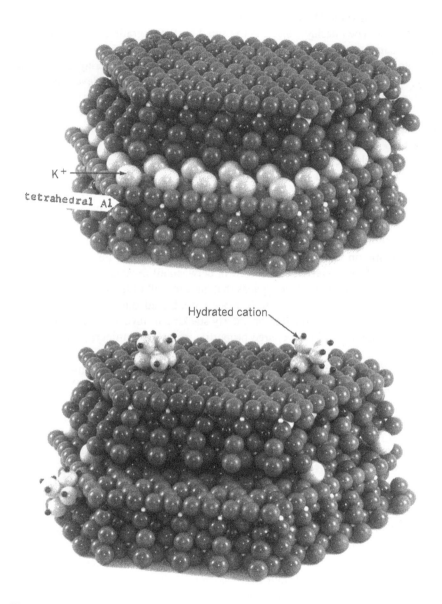

Figure 2.6 Models illustrating the weathering of mica to vermiculite. (a, top) Model of muscovite with two layers slightly offset to show the location of the interlayer unhydrated potassium ions that neutralize the negative charge in the tetrahedral sheet. 2.6 Models illustrating the weathering of mica to vermiculite. (b, bottom) Model of partially weathered mica with two layers slightly offset showing the loss of much interlayer potassium and creation of cation exchange capacity. Three exchangeable hydrated cations are shown: two on planar surface by CEC_c and one on the edge by CEC_v. (c, opposite) Model of vermiculite with expanded interlayer space occupied by hydrated and exchangeable cations.

Hydrated
cation

Figure 2.6 (continued)

Table 2.1 Charge Characteristics of Some Materials Removed from Kenya Soils

Material	Cation exchange capacity, meq/100g			Anion exchange capacity, meq/100g
	Permanent	Variable	Total	
Montmorillonite	112	6	118	1
Vermiculite	85	0	85	0
Weathered mica	11	8	19	3
Kaolinite	1	3	4	2
Gibbsite	0	5	5	5
Goethite	0	4	4	4
Allophanic colloid	10	41	51	17

Source: Data on Kenya soils adapted from Sanchez, 1976.

and the low energy of hydration of K, resulting in easy loss of hydrated water when the K^+ are pulled into the wedge-shaped interlayer spaces. The fixation of K lowers the CEC and causes the expanded edges of layers to collapse. Potassium fixation in agricultural soils protects K from being lost by leaching. Ammonium also has a low hydration energy. It is fixed similarly to K and protected from loss by leaching.

In summary, mica and vermiculite are 2:1 layer phyllosilicates that represent end members of a limited weathering sequence. The intermediate members, partially weathered micas, have properties intermediate between mica and vermiculite. As the weathering of micas proceeds, there is a marked increase in the CEC_c from the loss of interlayer K. Complete loss of interlayer K results in expansion of the interlayer space and the transformation of mica to vermiculite. Vermiculite has high negative charge from isomorphous substitution of Al for Si in the tetrahedral sheets resulting in high CEC_c. There is a small amount of variable negative charge and CEC_v and a small amount of positive charge and AEC. The net charge is strongly negative. Mica, weathered mica, and vermiculite contribute significantly to the charge properties of soils throughout the world, owing to the ubiquitous distribution of mica in soils.

2.6.2 Smectites

Smectites are 2:1 layer-expanding minerals with significantly less permanent negative charge and CEC_c than vermiculite, and little or no ability to fix K. The most important soil smectites are dioctahedral and include montmorillonite, beidellite, and nontronite.

Much of the smectite in the Mollisols in the U.S. was inherited from the loessial parent material. It persists in these soils because smectite is stable in slightly acid, neutral, and alkaline environments. Smectite disappears from many humid region soils because it is unstable in these acid environments. Smectites are the dominant clay mineral in the cracking soils of the Vertisol order. The Vertisols of Texas and Alabama inherited much smectite from their marine parent material. Smectite in the Vertisols in Hawaii and on the Deccan Plateau in India was formed by the weathering of basaltic rocks (rich in Si and Mg) where rainfall is limited and there are distinct annual wet and dry seasons. An illustration of pedogenic and inherited smectite is shown in Figure 2.7.

Figure 2.7 Pedogenic montmorillonite is forming in the soils on the slopes from the weathering of basalt in Hawaii. Inherited montmorillonite occurs in the basin in the foreground from erosion. The expanding nature of smectites is shown by the large cracks. (From *Fundamentals of Soil Science*, 7th ed., H. D. Foth. © 1984. Reprinted by permission of John Wiley & Sons.)

Alteration of vermiculite into smectite occurs by dealumination. This is the replacement of some of the Al by Si in the tetrahedral sheet resulting in some loss of constant negative charge. This is a likely means of formation of beidellite that has a significant amount of tetrahedral Al. The source of constant negative charge from isomorphous substitution (Al^{3+} for Si^{4+}) in tetrahedral sheets is reflected in the idealized formula for beidellite:

$$[Si_{3.5}Al_{0.5}][Al_2]O_{10}(OH)_2Ca_{0.25}$$

Ideally, the CEC_c is 135.2 cmol kg^{-1}. The CEC_c is shown to be neutralized by exchangeable Ca, $Ca_{0.25}$. In soils, however, all cations in the soil solution compete for neutralization of the cation exchange sites.

Most of the smectite in soils is montmorillonite. Its formation and persistence in soil is favored by an environment with limited leaching and a soil solution enriched with Si and Mg. There is some isomorphous substitution of Mg^{+2} for Al^{+3} in the octahedral sheet. In arid and subhumid regions, limited leaching produces conditions favorable for montmorillonite formation and its preservation.

Soil solutions enriched with Si and Mg in humid regions result from restricted drainage and inhibited removal of the weathering products by leaching. Thus, soils in low parts of a landscape with restricted drainage may contain smectite and well-drained soils on the surrounding slopes may be more acid and have a mineral fraction dominated by other clay minerals. The result is soils with widely different clay-mineral suites and soil fertility characteristics within a given field or region.

The idealized formula for montmorillonite is:

$$[Si_4][Al_{1.5}Mg_{0.5}]O_{10}(OH)_2Ca_{0.25}$$

Isomorphous substitution creates a high constant negative charge, ideally the CEC_t is 135.5 cmol kg^{-1}. The CEC_t is commonly reported to be about 80 to 120 $cmol_c$ kg^{-1}. About 10% of the total charge, CEC_t, is pH dependent. The charge distribution for a montmorillonite specimen is given in Table 2.1.

The constant charge in the octahedral sheet is farther away from the interlayer space, as compared to vermiculite. Adjacent 2:1 layers are more expandable because the attraction of adjacent layers to interlayer cations is much less and the hydrated interlayer cations can more easily push the layers apart. The interlayer distance of the smectites is highly variable, depending on water content, and may be two or more times greater than that of vermiculite. The tendency to fix ions in the interlayer space is greatly reduced. Thus, montmorillonite has great capacity to expand with wetting and drying and many layers exist as single entities, as compared to many layers making up a single clay particle. Montmorillonite has the largest specific surface of common soil clays, being 600 to 800 m^2 gm^{-1}. The model of smectite in Figure 2.8 illustrates the great expandability of the 2:1 layers resulting in (1) free access to the interlayer space by hydrated exchangeable cations, (2) inability to trap or fix ions in the interlayer space, and (3) access to all of the surface of the 2:1 layers for adsorption, including exchangeable cations on both inner and outer planar surfaces.

Figure 2.8 Model of smectite showing a wide interlayer space occupied by hydrated exchangeable cation. The interlayer space increases with wetting and decreases with drying.

The weathering of Fe-rich minerals in an environment conducive to smectite formation results in formation of nontronite with the following idealized formula:

$$[Si_{3.5}Al_{0.5}][Fe_2]O_{10}(OH)_2 Ca_{0.25}$$

Nontronite can also be inherited from the parent material. It is less abundant in soils than montmorillonite or beidellite.

In summary, smectites differ from vermiculites primarily in having (1) less constant negative charge and (2) lattice charge originating farther from the interlayer plane, resulting in greater expansion of layers and inability to fix K. Both vermiculite and smectites have very high CEC_c and low CEC_v and AEC.

2.6.3 Hydroxy-Aluminum Interlayered Vermiculite and Smectite

The formation and continued existence of vermiculite and smectite in soils are dependent on a low or only moderately acid weathering and leaching environment. The pH is about 6 and limited leaching results a soil solution enriched with Si. When the pH drops below 6, the amount of Al^{3+} available for neogenesis, relative to Si, increases because of the preferential loss of Si by leaching. This limits the amount of Si available to form 2:1 layers and Al polymerizes and forms large hydroxy-Al cations with positive charge. Aluminum oxide polymerization and formation of hydroxy-Al is enhanced near the surfaces of clay particles.

The large hydroxy-Al molecules are positively charged because of incomplete coordination of Al with OH in an acid weathering environment and their composition is related to soil pH. The Al is ideally in six-coordination with the sum of the OH$^-$ and H$_2$O equal to six. The Al to OH ratio of hydroxy-Al is commonly in the range 2.5 to 2.7, resulting in a positive charge. The smallest polymer believed to be nonexchangeable has a six-membered ring structure and the formula of Al$_6$(OH)$_{15}{}^{3+}$.

The positively charged hydroxy-Al, being strongly adsorbed, is not exchangeable and adsorption reduces the CEC. Adsorbed interlayer hydroxy-Al acts as a prop in the interlayer space of 2:1 layer expanding minerals, causing the layers to remain permanently expanded. Ion fixation in the interlayer space is reduced. The clay becomes nonexpanding due to the attraction of the positively charged hydroxy-Al interlayer and the negatively charged 2:1 layers (Figure 2.9).

Hydroxy–Al

Hydrated, exchangeable cation

Hydroxy–Al

Figure 2.9 Model of hydroxy-Al interlayered 2:1 expanding minerals, such as vermiculite and smectite. Adsorption of hydroxy-Al in the interlayer space reduces the CEC$_c$ and converts expanding clay to nonexpanding clay. Hydroxy-Al adsorption on planar surfaces also reduces the CEC$_c$.

Hydroxy-Al interlayering is favored by about 5.0 pH, release of significant Al by weathering, low organic matter content, and frequent wetting and drying of the soil. The hydroxy-Al occurs as islands first, then with time forms a continuous interlayer. The clays are then called 2:1:1 clays and they have greatly reduced CEC$_c$, compared to vermiculite and smectite. The CEC$_t$ ranges from nearly as much as vermiculite and smectite to as little as mica, depending on the extent of hydroxy-Al filling of the interlayer space. The CEC$_c$ is reduced more relatively than the CEC_v. Surface area also is greatly reduced. Along the edges of particles, exposed positively charged hydroxy-Al contributes to the AEC.

Hydroxy-Al interlayer clays give 2:1 layer phyllosilicates great weathering resistance and the interlayered clays are an important component of both moderately weathered and intensively weathered soils, including Alfisols, Ultisols, and some Oxisols.

2.6.4 Chlorites

Chlorites are an infrequent component of soil clays and are generally inherited from parent materials derived from metamorphic and igneous rocks. Chlorites are 2:1 layer minerals with the interlayer filled with a metal-hydroxide octahedral sheet that is positively charged. Adjacent negatively charged layers of the 2:1 structure strongly attract the positively charged hydroxide interlayer, resulting in a nonexpanding mineral.

Several different metals exist in the interlayer hydroxide sheet. The Mg–OH interlayer sheet contains Mg as the coordinating metal and has been called the brucite sheet. The Al–OH interlayer sheet has Al as the coordinating metal and has been called the gibbsite sheet. Chlorites weather rapidly in acid soils and tend to be readily transformed into vermiculitic or smectitic minerals.

Hydroxy-Al interlayer vermiculite (HIV) and smectite (HIS) have a structure similar to chlorite when the interlayer is continuous or complete. As a result, HIV and HIS are called secondary chlorite. Since their formation and abundance in soils differs greatly from primary chlorite, 2:1 clays with pedogenic continuous Al-hydroxy interlayers are referred to as hydroxyl-Al interlayer minerals, such as HIV and HIS, or as chloritized vermiculite and smectite.

2.6.5 Illite

The mica in soils inherited from parent material derived from granite weathering is largely muscovite and/or biotite, and partially weathered muscovite and/or biotite particles. Much mica inherited from argillaceous sediments and abundant in the till of the north-central U.S. was produced by *illitization* and is called illite.

The deposition of smectite in a basin, geologically speaking, is eventually transformed into shale, and with time undergoes alumination and repotassication under the right conditions to form illite as follows:

$$\text{smectite} + Al^{3+} + K^+ = \text{illite} + Si^{4+} \tag{2.7}$$

In the process there is isomorphous substitution of tetrahedral Si by Al which increases the lattice charge, replaces interlayer potassium, and becomes nonexpanding. The mineral, illite, is mostly dioctahedral and has less substituted Al than muscovite. Micas generally have low CEC_t. The CEC_t for illite is commonly given as 20 to 40 $cmol_c$ kg^{-1}. Some of this CEC is likely due to the presence of some interstratified smectite or vermiculite. The charge distribution data for weathered mica in Table 2.1 is from an illite specimen. The charge of partially weathered mica is importantly related to particle size and surface area.

Illite weathering in soils is comparable to the weathering of mica, including the loss of interlayer K and the formation of 2:1 layer expanding minerals. Such minerals also become chloritized in an acid weathering environment. Illite has a significant K content, being slightly lower than for muscovite. For practical purposes, illite can be considered similarly to micas in terms of nature and amount of charge and the release and fixation of K.

Illite has also been called hydrous mica, sericite, and clay mica. It is a common component of argillaceous sediments and rocks and, therefore, a common mineral in soils where parent materials develop from such sediments.

2.7 VARIABLE CHARGE 1:1 LAYER PHYLLOSILICATES

Kaolinite and halloysite are the most abundant 1:1 layer phyllosilicates in soils. Halloysite occurs in young soils that have developed in volcanic materials. Kaolinite is the most abundant phyllosilicate in intensively weathered soils. They both have similar structure and properties.

2.7.1 The 1:1 Layer Structure

The 1:1 layer structure results from the polymerization of one Si tetrahedral sheet and one Al octahedral sheet. Little, if any, isomorphous substitution occurs. The charge is nearly all variable and low. The basal plane of the tetrahedral sheet is all oxygen, as in the case of the 2:1 layers. The outer plane of the octahedral sheet is all hydroxyl. There is a relatively greater abundance of Al-coordinated hydroxyl exposed at the layer edges for development of variable charge, as compared to the 2:1 layer structure. The 1:1 layer is characterized by a lack of isomorphous substitution and low variable negative and positive charge. The ideal formula for the 1:1 structure is:

$$[Al_2][Si_2]O_5(OH)_4$$

2.7.2 Kaolinite

Kaolinite readily crystallizes from the ions released by the weathering of micas, feldspars, 2:1 and 2:1:1 layer clay minerals, and other aluminosilicate minerals. Preferential loss of Si relative to Al (desilication) in an acid leaching environment favors kaolinite formation. The Si:Al ratio decreases from about 2:1 to 1:1 during the formation of kaolinite from 2:1 layer clays. Equation 2.8 represents the neogenesis of kaolinite from orthoclase and the formation of soluble silica (silicic acid) and K^+

$$2KAlSi_3O_8 + 2H^+ + 9H_2O = \text{orthoclase}$$

$$Al_2Si_2O_5(OH)_4 + 4Si(OH)_4 + 2K^+ = \text{kaolinite}$$

$$(2.8)$$

The dissolution of the tetrahedral and octahedral sheets of 2:1 clays during acid weathering also provides silicic acid and Al for kaolinite formation:

$$2H_4SiO_4 + 2Al^{3+} + H_2O = Al_2Si_2O_5(OH)_4 + 6H^+ \qquad (2.9)$$

Kaolinite is commonly inherited from parent material derived from rocks and sediments composed of materials that originated from intensively weathered soils. It is one of the most abundant clay minerals in the Ultisols on the coastal plain in the southeastern U.S. The kaolinite was in the acid sediments from which the soils formed. Kaolinite is associated with intensive weathering and is a prominent clay mineral in Ultisols and Oxisols. A few Mollisols and Vertisols are kaolinitic, owing to inheritance of kaolinite from their parent material.

Kaolinite is nonexpanding and exists in soils as large clay-sized plates. It commonly occurs as particles large enough to be silt. The hydroxyls of the planar surfaces of the octahedral sheet electrostatically bond to the basal oxygen of the adjacent tetrahedral sheet. This bonding results in a nonexpanding clay with low specific surface. A model of kaolinite is shown Figure 2.10.

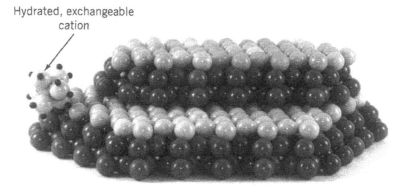

Hydrated, exchangeable cation

Figure 2.10 Model of two kaolinite layers slightly offset. The clay is nonexpanding and there is no interlayer space. Variable charge originates from protonation and deprotonation of edge-exposed hydroxyl. A hydrated exchangeable cation is adsorbed to a deprotonated hydroxyl on the particle edge.

There is uncertainty whether or not kaolinite may inherit some isomorphous substitution charge when formed by the alteration of 2:1 layer clay. Generally, isomorphous substitution in kaolinite is absent and the CEC_t is low. The charge is essentially all variable — arising from the deprotonation and protonation of exposed hydroxyl coordinated with edge-exposed Al. The charge depends on the purity of the clay and particle size. Many determinations on soil samples have included other clays as impurities. Values for CEC are commonly given in the range of 1 to 10 $cmol_c$ kg^{-1}. The charge distribution of a kaolinite specimen is given in Table 2.1.

The basal oxygen coordinated with Si^{4+} have one unsatisfied negative bond and protonate to form neutral and stable exposed OH. This OH is unreactive and unimportant in contributing variable charge to kaolinite. The middle and upper edge-exposed anions coordinated with Al are the source of variable charge. Normally, kaolinite exists in soils acid enough for the mineral to be net positively charged.

2.7.3 Halloysite

Halloysite is a 1:1 layer phyllosilicate similar to kaolinite, except the 1:1 layers are separated by a layer of water molecules when fully hydrated. Halloysite is usually formed from volcanic ash and glass. The clay is common in young soils derived from volcanic materials, such as the Andisols in the northwestern U.S., Alaska, and Hawaii. As with kaolinite, the clay has low pH-dependent charge and surface area.

2.8 VARIABLE CHARGE SHORT-RANGE ORDER ALUMINOSILICATES

The 1:1 and 2:1 layer phyllosilicates discussed thus far have atoms packed in a more or less orderly manner and form crystals, or particles, with distinct properties, such as shape, etc. They are long-range order minerals. The alumino-silicates, allophane and imogolite, are short-order range minerals. The bonding between Si, Al, O, and OH occurs only over a short distance. Their structure is somewhere between an amorphous precipitate and a highly crystalline material. Allophane occurs as spheres that clump together to form aggregates and imogolite occurs as tubes arranged into bundles. Allophane and imogolite are formed by the weathering of volcanic ash and other amorphous volcanic materials. They are the dominate minerals of soils developed from volcanic ash, such as Andisols (Figure 2.11).

These minerals have very large surface area. They develop variable charge from protonation and deprotonation of OH. Cation exchange capacity values of 10 to 40 $cmol_c$ kg^{-1} have been reported at pH 7. The charge distribution of an allophane specimen is given in Table 2.1.

Allophane and imogolite have a short life under leaching conditions. Over time, allophane and imogolite seem to transform to halloysite if the weathering environment favors resilication and to gibbsite if the weathering environment favors desilication. They form clays having greater crystallinity and greatly reduced surface area, with greatly reduced CEC_v and AEC.

These minerals complex with organic-matter and their presence gives many soils a high organic-matter content and black color.

Figure 2.11 Landscape on the North Island, New Zealand, where soils are mainly Andisols that developed from volcanic materials. Sheep grazing is the dominant land use.

2.9 VARIABLE CHARGE OXIDE MINERALS

In the early weathering stages, the soil solution usually receives considerable Si, Al, and Fe. Much of the Si and Al goes into the formation of aluminosilicate clay minerals. The released iron forms several stable crystalline minerals. These minerals accumulate in soils and coat sand, silt, and clay particles. Iron has an important role in determining soil color, since iron minerals have high pigmenting power. While Al and Si play an important role in aluminosilicate formation, the iron released ends up mainly in iron oxide minerals that accumulate in soils with time. The iron minerals, mainly oxides, gradually coat the surfaces of other mineral particles and alter their charge properties.

Desilication approaches or reaches completion in intensively weathered Oxisols and Ultisols. Thus, iron and aluminum minerals increase in abundance with time and make up an increasing part of the soil mass. The clay fraction may become dominated with iron and aluminum oxides along with varying amounts of kaolinite. Soils with abundant gibbsite tend to have a significant amount of kaolinite because these two clays tend to form near the end of the weathering sequence.

2.9.1   Gibbsite

There are several crystalline forms of Al oxide in soils; gibbsite is the one of great importance. With time, an acid weathering and leaching regime in humid region soils causes continued desilication and disintegration of the tetrahedral sheets of kaolinite, thus transforming kaolinite into gibbsite:

$$Si_2Al_2O_5(OH)_4 + 5H_2O = 2Al(OH)_3 + 2Si(OH)_4 \text{ gibbsite} \qquad (2.10)$$

The gibbsite structure is similar to the hydroxy-Al interlayer sheet with Al in six-coordination with OH (see Figure 2.4). The mineral is dioctahedral with Al occupying two of every three octahedral cation positions. Gibbsite is crystalline and tends to occur in hexagonal plate-shaped particles with layers weakly held together. Particles are plate-shaped with the longest dimension being about 20 times greater than their thickness.

At the edges of gibbsite, exposed OH deprotonate at high pH to create negative charge and CEC_v and protonate at low pH to create positive charge and AEC. It has been reported that in pH range of 8 to 9.2, the CEC equals the AEC and the net charge is zero. Thus, in acid soils where gibbsite is abundant, the CEC of gibbsite is very low and less than the AEC and the net charge is positive. The charge distribution of a gibbsite specimen at pH 8.2 is given in Table 2.1.

2.9.2 Iron Oxides

Iron oxides are the most abundant of the metallic oxides in soils. They are mostly crystalline. However, they vary in degree of crystallinity and have decreasing surface area and reactivity with increasing degree of crystallinity. Goethite (FeOOH) tends to give soils yellow color and hematite (Fe_2O_3) is red colored. Iron in goethite is in six-coordination with O and OH and in hematite the iron is in six-coordination with O. When placed in solution, iron oxide minerals without a hydroxylated surface develop a hydroxylated surface. Surface OH are reactive and develop charge. A zero net charge of most of the iron oxides is in the pH range of 7 to 9.

Aluminum is a common isomorphous substitution with little or no effect on charge properties. The substitution of Ti^{4+} for Fe^{3+} has been reported.

Aluminum and Fe oxide minerals form complexes with organic matter and decrease the CEC of the SOM.

2.10 CHARGE CHARACTERISTICS OF SOILS

Some young soils develop from sediments or rocks composed of the weathering products of intensively weathered soils. These young soils have the mineralogical composition of intensively weathered soils. Many of the world's soils, however, fit the pattern that the mineralogical composition changes with time, according to the mineral weathering sequence concept. As a consequence, the charge properties of minimally and moderately weathered mineral soils tend to be greatly influenced by 2:1 layer phyllosilicate minerals. Intensively weathered soils are greatly influenced by 1:1 layer phyllosilicate and oxide minerals. Organic matter contributes variable negative charge to all soils and organic soils are variable charge soils. A charge classification of materials and soils, developed by Uehara and Gillman, is based on the extent and type of charge of the soil's specific surface as:

60% or more constant charge   constant charge
60% or more variable charge   variable charge
other charge properties   mixed charge

Five soil orders are clearly variable or constant charge soils, including Vertisols as constant charge soils and Andisols, Histosols, Oxisols, and Spodosols as variable charge soils. Estimates of world distribution are 29, 32, and 39%, respectively, for variable, constant and mixed charge soils as given in Table 2.2. For the tropics, about 60% of the soils are variable charge, and for the temperate regions about 45% are mixed and 45% are constant charge.

Table 2.2 World Distribution of Soils with Variable, Mixed, and Permanent Charge Minerals

Surface Charge	Tropical zone, %	Temperate zone, %	World, %
Variable	60	10	29
Mixed	30	45	39
Permanent	10	45	32

Source: Uehara and Gillman, 1981.

2.11 SUMMARY

The charge in soils plays an important role in the retention of exchangeable anions and cations, potassium and ammonium fixation, acidity and basicity, and interaction with other components including biocides. The charge characteristics of soils are the net effect of the organic and mineral components. Except for a very few, soil horizons are net negatively charged.

Four major groups of soils are based on charge characteristics:

1. Mineral soils with moderate to high constant charge from 2:1 layer phyllosilicate clay minerals. Most of these soils are in the orders Alfisols, Aridisols, Entisols, Inceptisols, Mollisols, and Vertisols.
2. Mineral soils with low variable charge from low activity crystalline clay minerals (1:1 layer phyllosilicate and oxide clay minerals) that are characteristic of Oxisols and Ultisols. Entisols and Inceptisols that occur in association with Ultisols and Oxisols will likely have variable charge characteristics.
3. Mineral soils with low to high variable charge from short range-order and amorphous aluminosilicate minerals and organic matter; the Andisols and Spodosols.
4. Organic soils (Histosols) with high variable negative charge from humus.

Each of the four major groups of soils have unique properties and kinds of response to management practices. This requires different kinds of soil fertility management programs to make and keep the soils productive.

REFERENCES

Allen, B. L. and D. S. Fanning. 1983. Composition and Soil Genesis. In *Pedogensis and Soil Taxonomy,* Vol. 1. L. P. Wilding, et al., Eds. Elsevier, Amsterdam.

Allen, B. L. and B. F. Hajek. 1989. Mineral Occurrence in Soil Environments. In *Minerals in the Soil Environment.* 2nd ed. Soil Sci. Soc. Am., Madison, WI.

Barnhisel, R. I. and P. M. Bertsch. 1989. Chlorites and Hydroxy Interlayered Vermiculite and Smectite. In *Minerals in the Soil Environment.* 2nd ed. Soil Sci. Soc. Am., Madison, WI.

Barrow, N. J. 1985. Reactions of Anions and Cations with Variable Charge Soils. *Advances in Agronomy.* 38:183–228.

Bohn, H. L., B. L. McNeal and G. A. O'Connor. 1985. *Soil Chemistry.* 2nd ed. John Wiley & Sons, New York.

Borchardt, G. 1989. Smectites. In *Minerals in the Soil Environment.* 2nd ed. Soil Sci. Soc. Am., Madison, WI.

Broadbent, F. E. 1954. Modification of Chemical Properties of Straw During Decomposition. *Soil Sci. Soc. Am. Proc.* 18:165–169.

Dixon, J. B. 1989. Kaolin and Serpentine Group Minerals. In *Minerals in the Soil Environment.* 2nd ed. Soil Sci. Soc. Am., Madison, WI.

Douglas, L. A. 1989. Vermiculites. In *Minerals in the Soil Environment.* 2nd ed. Soil Sci. Soc. Am., Madison, WI.

Fanning, D. S., V. Z. Keramidas and M. A. El-Desoky. 1989. Micas. In *Minerals in the Soil Environment.* 2nd ed. Soil Sci. Soc. Am., Madison, WI.

Foth, H.D. 1984. *Fundamentals of Soil Science.* 7th ed. John Wiley & Sons, New York.

Foth, H.D. 1990. *Fundamentals of Soil Science.* 8th ed. John Wiley & Sons, New York.

Helling, C. S., G. Chesters and R. B. Corey. 1964. Contribution of Organic Matter and Clay to Soil Cation-Exchange Capacity as Affected by pH of the Saturating Solution. *Soil Sci. Soc. Am. J.* 28:517–520.

Hsu, P. H. 1989. Aluminum Oxides and Oxhydroxides. In *Minerals in the Soil Environment.* 2nd ed. Soil Sci. Soc. Am., Madison, WI.

Keng, J. C. W. and G. Uehara. 1973. Chemistry, Mineralogy, and Taxonomy of Oxisols and Ultisols. *Soil and Crop Sci. Soc. Florida.* 33:119–126.

Lucas, R. E. 1982. Organic Soils (Histosols). *Research Report 435.* Mich. Agr. Exp. Sta., East Lansing, MI.

Mehlich, A. 1981. Charge Properties in Relation to Sorption and Desorption of Selected Cations and Anions. In *Chemistry in the Soil Environment.* Am. Soc. Agronomy Special Pub. No. 40, pp. 44–75.

Sanchez, P. A. 1976. *Properties and Management of Soils in the Tropics.* John Wiley & Sons, New York.

Schwertmann, U. and R. M. Taylor. 1989. Iron Oxides. In *Minerals in the Soil Environment.* 2nd ed. Soil Sci. Soc. Am., Madison, WI.

Soil Survey Staff. 1975. Soil Taxonomy. *USDA Agr. Handbook 436.* Washington, D.C.

Sposito, G. 1989. *The Chemistry of Soils.* Oxford University Press, New York.

Stevenson, F. J. 1986. *Cycles of Soil.* John Wiley & Sons, New York.

Tan, K. H. 1993. *Principles of Soil Chemistry.* 2nd ed. Marcel Dekker, New York.

Tessens, E. and S. Sauyah. 1982. Positive Permanent Charge in Oxisols. *Soil Sci. Soc. Am. J.* 46:1103–1106.

Thomas, G. W. 1977. Historical Developments in Soil Chemistry: Ion Exchange. *Soil Sci. Soc. Am. J.* 41:230–238.

Thompson, H. S. 1850. On the Absorbent Power of Soils. *J. Royal Agric. Soc. Eng.* 11:68–74.

Uehara, G. and G. Gillman. 1981. *The Mineralogy, Chemistry, and Physics of Tropical Soils with Variable Charge Clays.* Westview Press, Boulder, CO.

Wada, K. 1989. Allophane and Imogolite. In *Minerals in the Soil Environment.* 2nd ed. Soil Sci. Soc. Am., Madison, WI.

CHAPTER 3

Ion Adsorption, Exchange, and Fixation

The charge in soils attracts and adsorbs ions from the soil solution. The ions are mainly hydrated, in motion, and exist at or near charged surfaces. Some ions are weakly or *nonspecifically* adsorbed mainly by electrostatic forces. Nonspecifically adsorbed ions are removed from soils by leaching with solutions containing an appropriate index ion for determining the cation and anion exchange capacity. Thus, nonspecifically adsorbed ions are equivalent to exchangeable ions and are available for plant uptake.

Some ions are strongly attracted mainly by covalent bonding and are *specifically* adsorbed. In general, specifically adsorbed ions are nonexchangeable, fixed and unavailable to plants and microorganisms, and not likely to leach to the groundwater.

Nonspecific and specific adsorption depend on the exchanger and the ion. For example, H$^+$ may be nonspecifically adsorbed onto clay and specifically adsorbed onto humus. Potassium is nonspecifically adsorbed onto a wide variety of negatively charged organic and mineral surfaces. It is specifically adsorbed onto micaceous minerals by negative charge resulting from isomorphous substitution.

The nature and extent of ion adsorption, exchange, and fixation largely determine the size and composition of the ion pool of the soil solution and affect plant growth and environmental pollution.

3.1 EXCHANGEABLE ION DISTRIBUTION NEAR CHARGED SURFACES

Exchangeable ions in dry soils are adsorbed directly onto charged surfaces. When dry soil is wetted, polar water molecules are attracted by both the ions and charged surfaces, forming hydrated ions and charged particle surfaces with water films. Some ions diffuse away from the charged surfaces. In addition, some of the electrostatic force is localized and some is nonlocalized, which also contributes to some exchangeable ions being drawn close to the charged surfaces and some ions being further away as shown in Figure 3.1.

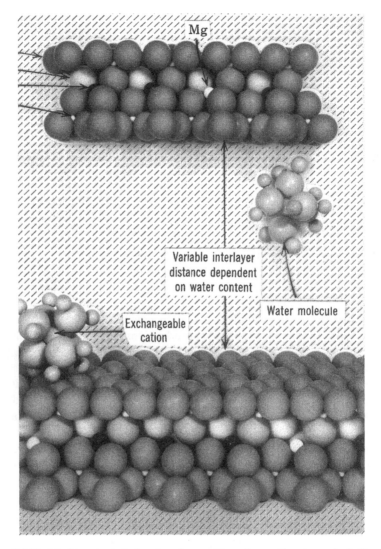

Figure 3.1 Model of two hydrated and exchangeable cations with primary hydration shells adsorbed on the planar surface by the negative charge owing to isomorphous substitution in a layer of 2:1 clay. The cation on the right is within the ion swarm and not the bulk solution. The hydrated cation on the left rests directly on the planar surface; note, however, that the cation is separated from the charged surface by water molecules of the primary hydration shell. (From *Fundamentals of Soil Science*, 7th ed., H. D. Foth. © 1984. Reprinted by permission of John Wiley & Sons.)

3.1.1 Ion Distribution Near a Negatively Charged Surface

The charged surfaces in soils are net negatively or positively charged or are at the zero point of charge (ZPC). Net positively charged surfaces preferentially attract and adsorb anions and repel cations. Most soil particle surfaces are net negatively charged, and thus, they preferentially adsorb cations and repel anions.

The strength of charge decreases rapidly within the soil solution with increasing distance from a charged surface. Diffusion causes some cations to move away from charged surfaces. Anions are repelled from a negatively charged surface. This produces a distribution of cations and anions near a negatively charged surface as shown in Figure 3.2.

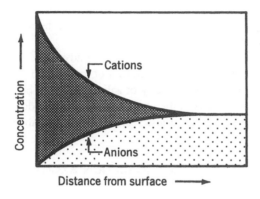

Figure 3.2 Distribution of cations and anions near the vicinity of a negatively charged surface.

The adsorbed ions make up the ion swarm. Beyond the ion swarm is the outer part of the soil solution (the bulk solution) where ions are unaffected by charged surfaces. In the outer soil solution, the concentration of cations is less than in the soil solution and that of the anions is more than the concentration in the ion swarm. The total charge of the cations in the ion swarm is equal to the negative charge of the surface plus the charge of the anions. The net charge of the system is zero with the negative charge of the cation exchange capacity (CEC) plus the anions, counterbalancing the positive charge of the cations and the anion exchange capacity (AEC).

There is no sharp line of demarcation between cations adsorbed close to charged surfaces and cations farther away, in the ion swarm. The cations are in motion and are constantly exchanging positions with each other. The cations under the influence of the charged surface tend to be retained when surplus water moves through the soil, while the cations in the outer soil solution are leached out of the soil. Factors that concentrate cations close to negatively charged surfaces include surface charge resulting from isomorphous substitution in tetrahedral sheets as compared to octahedral sheets, high surface charge density, an increase in the valence of cations, an increase in total electrolyte concentration, and a decrease in water content.

3.1.2 Ions in the Bulk Solution

The ions in the outer or bulk solution are free to move wherever the water moves. These ions move laterally to root surfaces by mass flow and diffusion and are leached downward in surplus water. The number of free ions in the bulk solution is usually small compared to the number of exchangeable ions, being of the order of 1% of the total ions of leached soils. There is an equilibrium between exchangeable and free ions. In fertile soils, the exchangeable ion pool acts as a buffer to maintain a supply of available nutrients in the outer soil solution for uptake by plants and microorganisms when ions are being removed by plant uptake and leaching.

Saline soils contain sufficient soluble salt to adversely affect the growth of most crops. They have a high concentration of ions beyond the ion swarm. An increase in the ion adsorption capacity of saline soils may be beneficial because it reduces the number of ions in the outer soil solution and decreases the osmotic pressure of the water resulting in less resistance in water absorption.

3.2 FACTORS AFFECTING THE CATION EXCHANGE CAPACITY

The capacity of a soil for the nonspecific adsorption of cations is the cation exchange capacity, CEC. Methods for determining the CEC are given in Chapter 2.

The CEC is produced from the amount and type of negative charge. It is mainly a function of the amount and kind of clay and organic matter and soil pH. Some soils have sufficient clay minerals and/or micaceous minerals in the silt fraction to have a significant effect on the CEC.

3.2.1 Role of Clay and Organic Matter

Removal of the organic matter and clay from soils and measuring the CEC of each fraction separately may alter the nature of the clay and organic matter and their interaction. Commonly, the CEC of many soils is determined, along with several other soil parameters known to affect the CEC, and a multiple regression analysis is used to estimate the CEC of both the clay and organic matter and their relative contribution to the CEC of soils. The multiple regression prediction equation for CEC has the form:

$$Y = b_1 x_1 + b_2 x_2 \tag{3.1}$$

where Y is the predicted CEC, b_1 and b_2 are the regression coefficients for organic matter and clay, respectively, and x_1 and x_2 are the percentages of organic matter and clay, respectively. The regression coefficients b_1 and b_2 represent the CEC of organic matter and clay, respectively, in $cmol_c$ kg^{-1}.

The A and B horizons of 344 soils of three Chernozemic subgroups developed from a wide range of parent materials and textures on the Great Plains of western Canada were analyzed. The CEC was measured at pH 7. Prediction equations developed for five horizons of three soils are given in Table 3.1.

Table 3.1 Prediction Equations for the Cation Exchange Capacity of Chernozemic Soil Horizons

Subgroup and horizon	Number of samples	Regression equation*
Rego, A horizon	21	$Y = 2.12x_1 + 0.56x_2$
Orthic, A horizon	111	$Y = 2.3x_1 + 0.57x_2$
Orthic, B horizon	99	$Y = 2.86x_1 + 0.57x_2$
Eluviated, A horizon	68	$Y = 2.38x_1 + 0.55x_2$
Eluviated, B horizon	45	$Y = 2.97x_1 + 0.59x_2$

*x_1 and x_2 are percentages of organic matter and clay, respectively.
Source: St. Arnaud and Septon, 1972.

The mean CEC of the soil organic matter (SOM) in the three A horizons has a range of 214 to 238 and a mean of 225 $cmol_c$ kg^{-1}. The mean CEC of the clay in the A horizons is 56 $cmol_c$ kg^{-1}, with a range of 55 to 57. Assuming that the CEC of the clay is 56 and of the SOM is 225, an A horizon with 20% clay and 4% organic matter would have a predicted CEC of 20.2 $cmol_c$ kg^{-1}. About 55% of the CEC is due to the clay and 45% is due to the SOM. The clay has high constant charge and is representative of many minimally and moderately weathered soils. Such soils commonly have A horizons in which about half of the CEC is due to clay and about half is due to SOM.

The SOM in the B horizons has greater CEC than in the A horizons, averaging about 291 $cmol_c$ kg^{-1} compared to 225 $cmol_c$ kg^{-1} for the A horizons. This suggests that the SOM of the B horizons is older and more decayed. The similar CEC of the clay fraction in A and B horizons indicates similar mineralogy. A Bt horizon containing 30% clay having a CEC of 58 $cmol_c$ kg^{-1} and 1% organic matter with a CEC of 291 $cmol_c$ kg^{-1} would have an estimated CEC of 20.3. Thus, about 86% of the CEC is attributable to the clay and only 14% is attributable to SOM.

These soils represent a toposequence of increasing age of well-drained Chernozemic subgroup soils from the driest landscape position (Rego) to the most moist landscape position (Eluviated). The Rego soils, in this case, are youthful soils on slopes with A and C horizons, which are subject to erosion. The eluviated soils are older and subject to surface deposition of eroded material from surrounding slopes. There is increasing CEC of SOM in the sequence Rego-Orthic-Eluviated associated with increasing average age of the SOM.

As moderately weathered soils evolve and become intensively weathered, the contribution of negative charge from the clay for CEC decreases, while that of SOM remains about constant. Thus, an Oxisol A horizon containing 20% clay with a CEC of 8 $cmol_c$ kg^{-1} and containing 4% SOM, similar to the Canadian soils, would have an estimated CEC of about 13.5 $cmol_c$ kg^{-1} with one-third of the CEC due to clay and two-thirds due to SOM. As soils weather and the CEC of the clay decreases, the relative contribution of the clay to the CEC decreases.

3.2.2 Role of Silt

The allocation of variance for the CEC of five soil horizons is given in Table 3.2. Organic matter and clay accounted for about 80% or more of the variation in CEC for all but one horizon, the eluviated A horizon. A small but significant amount of the predicted CEC is due to the interaction of clay and organic matter.

Table 3.2 Allocation of Variance in Cation Exchange Capacity of Five Chernozemic Soil Horizons

Subgroup and horizon	Due to variation in content of organic matter and clay (%)	Due to interaction (%)	Due to unknown sources (%)
Rego, A horizon	89	4	7
Orthic, A horizon	79	<1	21
Orthic, B horizon	81	4	15
Eluviated, A horizon	54	3	43
Eluviated, B horizon	80	2	18

Source: St. Arnaud and Septon, 1972.

The 43% variation caused by unknown sources for the Eluviated A horizon is likely associated with the eluviation of clay which causes components other than clay to make a greater contribution to the CEC. The topographic position of the Eluviated soils typically results in greater silt content of the A horizons for soils of the toposequence. Silt is likely a greater contributor to the CEC of the Eluviated A horizons as compared to the other horizons. The authors postulated that 15% silt in the 2-20 µm size range (fine silt) can account for as much as 10% of the CEC.

Loess is the parent material for about 10% of the world's soils. Many of these soils are found on the plains of the U.S., Canada, Argentina, and in Eurasia. Many of the A horizons contain about 60 to 70% silt, 20 to 40% clay, and only a few percentages of sand. In these soils, the silt may contribute a significant amount to the soil's CEC. The coarse silt, fine silt, coarse clay, and fine clay fractions were removed from six loess-derived soils in Iowa. The organic matter in the fractions was destroyed and the CEC determined on the remaining mineral material. The CEC in $cmol_c$ kg^{-1} was 3.6, 19.9, 51.6, and 60.1 for the coarse silt, fine silt, coarse clay, and fine clay, respectively. Averaging the CEC for the silt and clay fractions and assuming soil contained 30% clay and 70% silt, the estimated CEC for the soil is 25 $cmol_c$ kg^{-1} of which 33% is attributable to the silt.

A complicating factor in CEC studies is that silt-sized soil particles are sometimes aggregates largely of clay particles. This is especially the case for Oxisols.

3.3 CATION EXCHANGE CAPACITY OF SOILS

In a CEC study of 12,000 data sets representing soils from the continental U.S., Hawaii, Puerto Rico, and other countries, the content of clay and organic carbon were highly correlated with CEC (0.001 confidence level). The CEC was determined at pH 7. The regression equation which expresses the relationship between CEC and percent organic carbon (OC) and percent clay (C) for all soil horizons grouped together, is:

$$CEC = 3.2 + 3.67OC + 0.196C \qquad (3.2)$$

The regression equation accounts for 55% of CEC, coming from the organic carbon and clay. The mean CEC of the organic carbon is estimated to be 367 $cmol_c$ kg^{-1} and the mean CEC of the clay as 19.6 $cmol_c$ kg^{-1}. Organic matter is generally considered to contain 58% carbon. Thus, the mean CEC of the SOM is about 212 $cmol_c$ kg^{-1}.

Regression equations for estimating the CEC of the A, B, and C horizons are as follows:

$$CEC \text{ A horizon} = 1.6 + 3.26OC + 0.201C \qquad (3.3)$$

$$CEC \text{ B horizon} = 2.9 + 4.68OC + 0.198C \qquad (3.4)$$

$$CEC \text{ C horizon} = 2.9 + 4.34OC + 0.277C \qquad (3.5)$$

These equations also show greater CEC of organic matter, in the B and C horizons, as compared to A horizons. The higher regression coefficient for the C horizon suggests less weathered horizons with higher CEC clays.

The number of samples in the regression analysis for equations 3.2, 3.3, 3.4, and 3.5 were 5,535, 1,150, 3,553, and 832, respectively.

3.3.1 Mineral Soil Orders

Cation exchange capacity, in order of increasing CEC, and other data based on all horizons of the soil orders included in the analysis of the large USDA data base are given in Table 3.3. Several trends are evident.

1. Oxisols and Ultisols have the lowest CEC. The low CEC is related to the low CEC of the clay fraction 8.4 and 16.7 $cmol_c$ kg^{-1} for Oxisols and Ultisols, respectively.
2. Entisols, Alfisols, Spodosols, and Inceptisols have comparable CEC. In this group, the Spodosols have clay with the lowest CEC, 19.3 $cmol_c$ kg^{-1}, and organic carbon with the highest CEC, 412 $cmol_c$ kg^{-1}. Greater fulvic acid content of the organic matter is one explanation for the higher CEC of the organic fraction. Spodosols have much lower clay content but higher organic carbon

Table 3.3 Cation Exchange Capacity, Organic Carbon, Clay Content, and CEC Prediction Equations for Soil Orders Worldwide

Order	CEC (cmol$_c$/kg)	Organic carbon (%)	Clay (%)	Regression equations to estimate CEC	R^2
Oxisols	8.9	1.1	47.8	Y = 2.31 + 2.60OC + 0.084C	0.46
Ultisols	9.9	0.6	26.4	Y = 1.63 + 3.00OC + 0.167C	0.51
Entisols	14.2	0.7	16.4	Y = 2.15 + 2.36OC + 0.304C	0.78
Alfisols	15.2	0.6	24.7	Y = 3.26 + 2.12OC + 0.351C	0.48
Spodosols	15.5	2.6	6.1	Y = 3.36 + 4.12OC + 0.193C	0.78
Inceptisols	16.0	1.5	17.1	Y = 3.45 + 3.66OC + 0.243C	0.67
Aridisols	16.4	0.5	20.2		
Mollisols	20.6	0.9	26.3	Y = 4.81 + 2.88OC + 0.352C	0.63
Vertisols	38.0	0.8	51.4	Y = −4.29 + 1.46OC + 0.624C	0.67

Source: Manrique, Jones, and Dyke, 1991.

content, compared to the other orders in this group, resulting in similar CEC as for the other soils.

3. Aridisols are grouped mainly on the basis of aridity resulting in a very heterogeneous group of soils. As a result, organic carbon and clay are not good predictors of CEC and no prediction equation is given in Table 3.3.

4. Mollisols appear to have slightly more CEC, likely related to greater organic matter content and slightly more smectitic clay, compared to soils in the previously mentioned orders. The CEC of the clay, however, is significantly lower than that for the Chernozemic soils of western Canada.

5. Vertisols have the highest CEC, owing to high content of smectitic clay with mean CEC of 62.4 cmol$_c$ kg^{-1}. The CEC of the clay is lower than expected, probably because of the inclusion of data of some Vertisols with some kaolinitic clay in addition to smectite.

Variation in CEC attributed to organic carbon and clay ranged from 46% (R^2 = 0.46) for Oxisols to 78% (R^2 = 0.78) for Entisols and Spodosols.

Organic carbon was the largest contributor to the prediction of CEC for Spodosols and for Oxisols. This is associated with the very low CEC of the clay in Oxisols and the very low clay content in Spodosols.

There is no data for Andisols because the order was only recently established. About 10% of the soils in the Inceptisol order are Andepts and that is likely the case for data in Table 3.3. Most Andepts are now classified as Andisols. Such soils develop from volcanic materials and have short range-order minerals. Andisols have CEC quite comparable to Vertisols. However, they are characterized by moderate-to-high organic matter content and high activity aluminosilicate clay that is mainly short range-order.

3.3.2 Organic Soils

Humus is the major contributor to the CEC of organic soils or Histosols. The CEC is related to the degree of decomposition. Fibrists are fibrous Histosols and are the least decomposed or most peaty Histosols with a CEC of about 100 cmol$_c$

kg^{-1}. The Saprists, the most decomposed Histosols, have a CEC more nearly 200 cmol$_c$ kg^{-1}. Hemists are intermediate in degree of decomposition and CEC. A CEC of 141 cmol$_c$ kg^{-1} has been reported for Histosols from a study of a very large number of soils in a Department of Agriculture data bank.

Where Histosols are contaminated with mineral material, such as volcanic ash, the properties may be importantly affected by mineral components.

Histosols commonly have a CEC between 100 and 200 cmol$_c$ kg^{-1}. This does not mean, however, that plant roots have many times more CEC in a given volume of root environment as compared to most mineral soils. One kilogram of an organic soil may represent a volume that is several times larger than one kilogram of mineral soil. Thus, it is useful to compare the CEC on a weight and volume basis as shown in Table 3.4. The data in the table reflect both a greater CEC and bulk density of muck as compared to the peat. On a volume basis, about the same amount of lime is required for the loam soil as compared to the woody peat to make a similar change in calcium saturation. Muck soils, however, require about four times more lime than loamy soils per unit volume of soil.

Table 3.4 Comparison of CEC on a Weight and Volume Basis

Soil	Weight basis (cmol$_c$/kg)	Volume basis (cmol$_c$/L)
Loam, mineral soil	12	14
Sphagnum peat	100	8
Woody peat	90	14
Muck	200	60

Source: Lucas, 1982.

Muck has a CEC and a degree of decomposition comparable to Saparists; peats are comparable to Fibrists.

3.4 THE EFFECTIVE CATION EXCHANGE CAPACITY

The CEC of soils has traditionally been determined at pH of 7.0 or 8.2, depending on method. This allows valid comparisons of CEC between soils. The CEC determined at the soil's current, or natural pH, is called the effective CEC (ECEC). The ECEC more accurately reflects conditions encountered by plant roots and microorganisms, especially in acid soils.

A study of the contribution of clay and organic matter (organic carbon × 1.72) to the CEC over the pH range of 2.5 to 8 was made on the Ap horizons of 60 Mollisols and Alfisols in Wisconsin. The soils averaged 13.3% clay, which can be expected to consist mostly of 2:1 and 2:1:1 types, plus a minor amount of oxide clay existing as particle coatings. The SOM averaged 3.38%. The CEC of the soil increased from about 8 cmol$_c$ kg^{-1} at pH 3.5 to about 15 at pH 8.0 as shown in Figure 3.3.

The CEC of both the clay and organic carbon increase with the increasing pH. The CEC of the organic matter increases from 36 cmol$_c$ kg^{-1} at pH 2.5 to

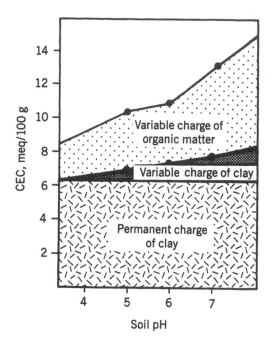

Figure 3.3 The average source of negative charge (CEC) in 60 Wisconsin soils. (Data from Helling, Chesters, and Corey, 1964.)

213 at pH 8 and the respective increase in the CEC of the clay is from 38 to 64 $cmol_c$ kg^{-1} (Figure 3.3). The percentage contribution to the CEC from organic matter increased from 19 to 45%, indicating á greater relative increase in CEC of organic matter, as compared to clay, with increasing pH. Extrapolating the data for organic carbon, which may not be valid, indicates that the CEC of the SOM approaches zero at pH 2.1.

Many soils in the world have a pH that is near neutral or alkaline, including many Aridisols, Mollisols, and Vertisols. In these soils the CEC and ECEC are the same for all practical purposes. Large relative differences between CEC and ECEC occur in acid Histosols, Oxisols, and Ultisols which are dominated by variable charge.

3.4.1 The Effective Cation Exchange Capacity Profile

The distribution with depth of clay and SOM and their charge characteristics give the soil its ECEC profile. Usually there is a decreasing amount of SOM with increasing soil depth. In immature soils, there tends to be a relatively constant clay content with increasing depth, and with time, clay moves downward from A and E horizons and accumulates in Bt horizons. The result is that A and Bt horizons may have similar CEC but relatively more of the CEC in A horizon is due to SOM, and relatively more of the CEC in the Bt horizon is due to clay.

3.4.2   Management of the Cation Exchange Capacity

In soils dominated by constant charge clays, the ECEC is generally high and, from a practical standpoint, difficult to alter. There are, however, relatively minor changes in the ECEC resulting from liming practices, but these are not considered important. More potential exists to manage the ECEC in intensively weathered acid soils with low ECEC (Figure 3.4).

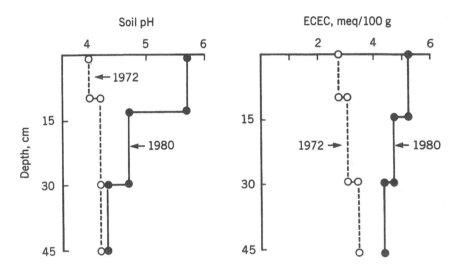

Figure 3.4 The effect of liming on soil pH and the effective cation exchange capacity of an Ultisol. (From Sanchez, Villachica, and Bandy, 1983. Used by permission of the Soil Society Science of America.)

About 4 $cmol_c$ kg^{-1} of ECEC is needed to have sufficient cation retention for cropping in humid regions. Liming of oxide charge dominated acid soils, such as Oxisols, can produce significantly important increases in the ECEC. Adding organic matter to increase the ECEC is very difficult. On the other hand, large losses of organic matter, as a result of land clearing, significantly decrease the ECEC.

3.5 CHARACTERISTICS OF CATION EXCHANGE REACTIONS

In general, cation exchange reactions are rapid, reversible, stoichiometric and obey the mass action law. The greater the concentration of any cation, and the more strongly the cation is attracted to the charged or exchanger surface, the more likely the cation will cause desorption of competing cations and be adsorbed in the ion swarm rather than the bulk solution.

3.5.1 Effects of Valence and Hydration

Valence is the major factor affecting the likelihood that a cation will be adsorbed. A trivalent ion like Al^{3+} is more strongly adsorbed than Ca^{2+} which is more strongly adsorbed than Na^+. It requires two Na ions to replace one Ca ion. Divalent cations are held with about twice the energy as monovalent cations. For cations of similar valence, the smaller cation will have the greater charge density per unit volume and, therefore, attract more water molecules and form a larger primary hydration shell. These water molecules cause the hydrated radius of a smaller cation to be larger than the hydrated radius of a larger cation. For this reason, Ca^{2+} is adsorbed with more energy than Mg^{2+} and K^+ more than Na^+. The general order of selectivity or replaceability of some important hydrated exchangeable cations is $Al^{3+} > Ca^{2+} > Mg^{2+} > K^+ = NH_4^+ > Na^+$.

It is difficult to place H in the series because it is weakly bonded to charged surfaces of clay minerals, but it may degrade clay minerals, releasing Al^{3+}. Thus, measurements of replacement power of H^+ may actually reflect Al^{3+} since it was released to solution by H^+. In addition, H^+ is strongly bonded to the acidic groups of organic matter. The result is that H is nonspecifically adsorbed to clays and preferentially, or specifically, adsorbed onto SOM.

The K^+ and NH_4^+ can readily shed their primary hydration shell and move directly to charged surfaces. When this occurs with micaceous minerals, the cations are pulled into voids formed by the earlier release of K^+ and become specifically adsorbed or fixed.

3.5.2 Complimentary Ion Effect

The exchange of one cation by another cation in the presence of a third or complimentary cation is easier if the complimentary cation has greater valence or energy of adsorption. During uptake of cations by roots, the root excretes H^+, which is monovalent. When the H^+ diffuses to the colloid surfaces, it is more likely to exchange K^+ than Ca^{2+} if the soil is acid and many trivalent Al^{3+} ions are adsorbed. By contrast, the H^+ would be less likely to exchange for K^+ than Ca^{2+} if the soil was alkaline and there was no exchangeable Al^{3+}, but much exchangeable Mg^{2+} and Na^+ as the complimentary cations. In effect, as soils become more acid and increase in exchangeable Al^{3+} below pH 5.5, the exchangeable K^+ is more likely to exist beyond the ion swarm in the bulk solution and to have greater mobility. The immediate effect is good because these soils tend to be low in exchangeable K^+; however, the long-term effect is greater leaching of K.

3.5.3 Dilution Effect

As soils dry, the quantity of soil solution decreases. This gives preference to adsorption of cations of lower valence or energy of adsorption. Conversely, the amount of water or soil solution increases and there is a dilution of cations and preference for adsorption of the more strongly adsorbed cations. As a consequence, soil drying tends to increase the adsorption of K relative to Ca and Mg, and increasing water content of soil increases the adsorption of Ca and Mg relative to K and Na.

3.6 THE EXCHANGEABLE CATIONS IN SOILS

Exchangeable cations represent the major available supply of many plant nutrients. Their absolute and relative amounts are important in affecting plant growth. The hydrolysis of Al and Na affects soil pH.

3.6.1 Kinds and Terminology

The dominant exchangeable cations are Al, H, Ca, Mg, K, and Na. In equation 2.1, XCa was used where X represented a negatively charged exchange surface and Ca represented exchangeable Ca. By switching the X to before Ca to produce XCa, the X now represents the fact that the Ca is exchangeable. Exchangeable cations are, therefore, indicated with an X placed before the ion. For example, XAl, means exchangeable Al and XMg means exchangeable Mg. Further, XAl..Na refers to the six dominant exchangeable cations (Al, H, Ca, Mg, K, Na) and their sum in cmol kg^{-1} is equated to the CEC. Other nutrients exist in soils as exchangeable cations, including Cu, Fe, Mn, Zn, Mn, and N as ammonium, but the amount is very small by comparison to the amount of XAl..Na. These other cations are not included in determination of the CEC if it is estimated by measuring cations exchanged by NH_4OAc.

3.6.2 The Exchangeable Cation Suite

The compliment of exchangeable cations is a function of soil parent material, the changes that occur during soil genesis, and the effects of soil management practices. Many parent materials and soils are calcareous and contain an abundance of $CaCO_3$ and smaller amounts of other salts. From the dissolution of these salts come the cations for adsorption. As salts dissolve, the cations entering solution are attracted to the ion exchange surfaces according to valence, hydrated radius, and mass action. The dominant exchangeable cations are Ca, Mg, K, and Na and their abundances generally parallel the energy of adsorption sequence, with Ca most abundant and Na or K the least abundant.

The leaching of carbonates produces noncalcareous soil with similar cation saturation as calcareous soil. Such soils have a pH of about 7 and are 100% saturated with Ca, Mg, K, and Na (XCa..Na). Additional leaching results in progressive loss of exchangeable Ca, Mg, K, and Na or XCa..Na. The loss of XCa..Na is accompanied by a comparable decrease in the ECEC. From pH 6 to 7, the soil remains 100% XCa..Na saturated. Thus, a pH change from 7 to 6 produces little change in the root environment and nutrient uptake.

Below pH 6, exchangeable H^+ and Al^{3+} increase in abundance, and the XCa..Na saturation decreases below 100%. There is a decreasing amount of available Ca, Mg, and K and increasing amount of XAl and XH; at pH near 4.5, they are nearly equal. At pH near 4, the CEC is nearly depleted of XCa..Na and 100% saturated with XAl and XH. This relationship between exchangeable cations and soil pH is the basis for calling XCa..Na basic cations and XH and XAl acidic cations. Exchangeable Mn increases in very acid soils because of increasing solubility of minerals containing Mn.

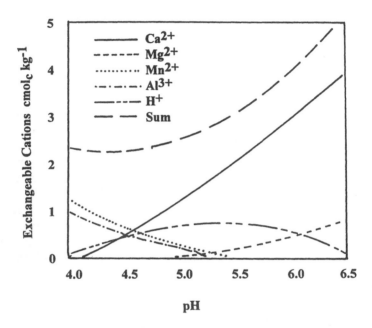

Figure 3.5 Exchangeable basic and acidic cations as a function of soil pH. (Data from Stumpe and Vlek, 1991. Used by permission of the Soil Science Society of America.)

The cation suite for acid soils ranges from 100% XCa..Na saturated near pH 7 to about 100% XAl and XH saturated at pH 4. Increasing acidity is associated with decreasing ECEC, decreasing supply of available essential plant nutrients, and an increase in the amount of solution H^+ and Al^{3+}.

For many years, basic cations were called exchangeable bases and the acidic cations called exchangeable acidity. Sometimes acid cation and base cation are used along with other terms. The usage here is acidic cations to refer to XAl and XH and basic cations to refer to XCa..Na. (For a discussion of the use of these terms, see letters by Binkley and others in the Soil Science Society of America Journal, page 1089-1091, 1987.)

3.6.3 Serpentine Soils

The mineralogical uniqueness of the parent material sometimes gives soils an unusual exchangeable cation suite. This happens with soils that have developed from serpentine parent materials. Serpentine minerals have a 1:1 structure similar to kaolinite and contain considerable isomorphous Mg and Fe. Weathering releases large amounts of Mg relative to the other cations. Montmorillonite is a common clay formed from serpentine weathering.

Four soils formed from serpentinite rock in California were found to have C and/or Cr horizons with average exchangeable cation ratios (cmol$_c$ kg^{-1}) of Ca:Mg:K:Na of 26:410:1:1. The A horizons were more weathered and leached, but even so, they had a high imbalance of XMg relative to XCa for plant growth with ratios of 61:174:3:1, respectively. Most plants cannot tolerate a cation suite that is particularly unbalanced relative to Ca and Mg. Calcium deficiencies are common in these soils. Land with serpentine soils have unusual flora and sparse plant growth and are referred to as *serpentine barrens*. Serpentine soil materials exposed by mining are very difficult to revegetate.

3.6.4 Saline and Sodic Soils

Saline soils contain salts more soluble than the carbonate salts found in calcareous soils. In arid and subhumid regions, water moves from one part of the landscape to another, depending on topography. In some parts of the landscape, runoff water collects in depressions and forms ponds. Subsequently, water evaporates and the salts that were in the water are deposited. The upward movement of water from a shallow water table to the soil surface may also furnish water to evaporate and deposit on or near the soil surface. Since the most soluble salts are moved to the greatest extent, soils may become *saline* from the infusion of soluble salt. Saline soils have sufficient soluble salt to impair plant growth.

The exchangeable cation suite of saline soils is highly variable, depending on the amount and kind of salts. Sodium salts are the most soluble and their preferential accumulation in time may result in a sodium adsorption ratio of 13 or more, or 15% or more XNa, and the formation of *sodic* soils. Soil structure tends to disintegrate in sodic soils because XNa is weakly adsorbed and is inefficient in neutralizing the negative charge. The diffuse ion swarm becomes more spread out. Dispersion of clays and humus greatly reduces permeability and tilth. The addition of gypsum, $CaSO_4 \cdot 2H_2O$, and leaching of the soil causes desorption and removal of XNa and an increase in the adsorption of XCa. The percentage XNa saturation decreases and the percentage of XCa increases. Eventually, the soil becomes nonsodic.

3.6.5 Summary Statement

In summary, we can state that:

1. Soils develop from a wide variety of parent materials and have great variation in degree of development. The exchangeable cation suite, however, shows much less variability. In some cases, the exchangeable cation suite is significantly affected by soil management practices.
2. The order of exchangeable cations is Ca > Mg > K > Na in most soils with a pH of 5.5 or more. Commonly, the amounts of XK and XNa are low. Exceptions occur in regard to sodic and serpentine soils. In sodic soils, Na may be the most

abundant exchangeable cation and Mg the most abundant exchangeable cation in serpentine soils.

3. Exchangeable aluminum appears below pH 5.5 and increases in abundance as soil pH declines. It is the most abundant exchangeable cation in many acid soils. Hydrogen is a minor exchangeable cation, even in acid soils.

3.7 EXCHANGEABLE CATIONS AS A SOURCE OF PLANT NUTRIENTS

In general, the greater the amount of an exchangeable cation the greater is its concentration in the bulk soil solution. As a consequence, the exchangeable cations represent an available supply of Ca, Mg, K, and Na for meeting the immediate needs of plants by root interception, mass flow, and diffusion.

3.7.1 Calculation of the Amount of Exchangeable Cations

The amount of certain exchangeable cations are routinely determined by soil testing labs, reported commonly as pounds per acre, and used as the basis for soil management decisions.

Problem: Calculate the amount of XCa per acre furrow slice for a soil that has an ECEC of 20 $cmol_c$ kg^{-1} and an XCa saturation percentage of 70.

1 mole of Ca^{2+} weighs 40 g
1 cmol of Ca^{2+} weighs 0.4 g
1 cmol of Ca^{2+} can saturate 2 cmol of CEC
Therefore, 0.2 g of Ca can saturate 1 cmol of CEC

14 cmol of CEC charge per kg are neutralized by XCa
14×0.2 g = 2.8 g XCa per kg of soil
2.8 kg XCa per 1,000 kg = 2.8 pounds XCa per 1,000 pounds of soil

An acre furrow slice is assumed to weigh 2,000,000 pounds oven dry; therefore:
2.8 pounds XCa/1,000 pounds soil = X pounds of XCa/2,000,000 pounds of soil

X = 5,600 pounds of XCa per acre furrow slice

3.7.2 Exchangeable Cations versus Plant Needs

A calcareous C horizon developed from loess in Iowa is reported to be saturated 67% by Ca, 30% by Mg, 2% by K, and 1% by Na; consider the adequacy of such soil for supplying the annual needs of crops.

The data in Table 3.5 shows that a plow layer of soil about 7 or 8 inches (20 cm) thick, would contain enough exchangeable Ca to meet the needs of productive crops for 86 years. For Mg and K, the supply would last 38 and 1.2 years, respectively. The soil can be expected to supply the Ca and Mg needs for many

Table 3.5 Amounts of Exchangeable Cations versus Plant Needs

Cation	Furrow slice		Supply, years*
	pounds/acre	kg/ha	
Ca	4,280	4,798	86
Mg	1,152	1,291	38
K	234	262	1.2

* Based on an average value of the annual uptake of productive crops and data for Ida silt loam, as reported in Soil Survey Investigations Report 3, USDA, 1966.

years because weathering constantly releases additional Ca and Mg. On the other hand, the large annual demand of K and the low exchangeable supply relative to annual needs means that K fertilizer is likely to be needed each year to produce high yields. The very small amounts of exchangeable Fe, Mn, Cu, and Zn are also important in meeting the nutrition needs of plants and microorganisms. Thus, it is difficult to overemphasize the importance of there being exchangeable cations in soils to serve as a source of available plant nutrients.

3.8 SPECIFIC CATION ADSORPTION

Strong attraction and specific ion adsorption involves ionic as well as covalent bonding. Such ions are considered fixed or unavailable and have very limited mobility in soils and limited threat of groundwater pollution.

The potassium ion readily loses hydration water molecules, allowing the K^+ to move directly to the surface of micaceous minerals where strong charge from isomorphous substitution in the tetrahedral sheet results in specific adsorption and fixation in cavities formed by six-member rings of oxygen ions (see Figure 2.6). Ammonium is similarly fixed.

Hydroxy-Al cations vary in size and charge. Some hydroxy-Al cations are exchangeable and some are specifically adsorbed onto clay minerals and SOM, resulting in reduced CEC. Interlayer hydroxy-Al is nonexchangeable, but is removable with appropriate treatment. Specific adsorption may also occur for Cu, Mn, and Zn.

3.9 ANION ADSORPTION AND EXCHANGE

Positive charge accounts for the adsorption and exchange of anions and the capacity of a soil, or soil component, to adsorb anions is the anion exchange capacity (AEC). Organic matter contributes to the CEC but does not contribute significantly to the AEC of soils. The positive charge in soils is related to the nature of the clay minerals and soil pH. The effective AEC (EAEC) is the anion exchange capacity at the soils current pH.

3.9.1   Nonspecific Anion Exchange

The nonspecific adsorption of anions parallels nonspecific cation adsorption. Mass action and strength of attraction largely control adsorption of anions such as Cl^- and NO_3^-. These anions are hydrated and exist in the ion swarm or as weakly adsorbed to charged surfaces. They are separated from charged surfaces by their hydrated shell. These anions are readily leached out of humid region soils. But they may be retained by very acid humid tropical soils.

3.9.2   Specific Anion Adsorption

Preferential or specific anion adsorption results through ligand exchange. Adsorption results when strongly attracted anions penetrate the ion swarm next to the exchange surface, even if predominately negatively charged, and penetrate the coordination sphere of a structural cation, like Al or Fe. The exchange is facilitated by increasing acidity and increasing abundance of positive charge. The adsorption of B, SO_4, and Mo are known to be more strongly adsorbed by amorphous Fe and Al hydroxides than for the more crystalline minerals. Phosphate and borate are specifically adsorbed. An example of phosphate adsorption is

$$-Fe-OH_2^+ + H_2PO_4^- = -Fe-O-PO(OH)_2 + H_2O \qquad (3.6)$$

There is a complete exchange of the water ligand from the Fe by the phosphate ion resulting in strong bonding between the iron and an oxygen of the phosphate ion. The ion is not separated from the adsorption site by a hydration shell and forms an inner sphere complex. Note that there has been a loss of an anion exchange site. In addition, the specific adsorption of P may occur by displacement of OH^- from Fe or Al oxide/hydroxide surfaces that are neutral. These reactions have also been called phosphorus fixation.

Sulfate adsorption is both nonspecifically and specifically adsorbed or is adsorbed intermediately between nitrate and chloride and the phosphate ion. Many soils are receiving precipitation with increasing amounts of sulfate from the burning of fossil fuels. The sulfate may contribute to increased soil acidity if the sulfate resulted from oxidation of reduced S. As a result, protonation of iron and aluminum oxides increases the soil's ability to adsorb sulfate.

3.10 SOILS WITH EQUAL AMOUNTS OF ANION AND CATION EXCHANGE CAPACITY

At the pH of the ZPC of soil, the negative and positive charge are equal (ECEC equals EAEC), and the net charge is zero. Soils with ion exchange surfaces dominated by 2:1 clay minerals and organic matter have much more negative than positive charge throughout the pH range relevant for crop production.

A ZPC in a soil is possible if the negative charge is low and conditions favor the development of positive charge. Conditions that contribute to a net zero or a positive charge soil are (1) low organic matter content, (2) low CEC variable-charge clay minerals, and (3) acidity or low pH. A net positive charge occurs in some subsoil horizons of extremely weathered Oxisols. These soil horizons preferentially adsorb anions, compared to cations. The EAEC and ECEC are both very low. In such Oxisols, the A horizons have a net negative charge owing to SOM, while the subsoils (oxic horizons) have a net positive charge.

Oxisols with net positive charged subsoil layers are in the anionic subgroups. For example, an Anionic Acrudox is an Oxisol with a udic moisture region that is extremely weathered, and has a soil layer that is net positively charged or has a net charge of zero.

REFERENCES

Barrow, N. J. 1985. Reactions of Anions and Cations with Variable Charge Soils. *Advances in Agronomy*. 38:183–228.

Binkley, D. 1987. Use of the Terms "Base Cation" and "Base Saturation" Should be Discouraged. *Soil Sci. Soc. Am. J.* 51:1089–1090.

Bohn, H. L., B. L. McNeal and G. A. O'Connor. 1985. *Soil Chemistry*. 2nd ed. John Wiley & Sons, New York.

Foth, H.D. 1984. *Fundamentals of Soil Science*. 7th ed. John Wiley & Sons, New York.

Foth, H. D. 1990. *Fundamentals of Soil Science*. 8th ed. John Wiley & Sons, New York.

Helling,C. S., G. Chesters and R. B. Corey. 1964. Contribution of Organic Matter and Clay to Soil Cation-Exchange Capacity as Affected by the pH of the Saturating Solution. *Soil Sci. Soc. Am. J.* 28:517–520.

Lucas, R. E. 1982. Organic Soils (Histosols). *Research Report 435*. Mich. Agr. Exp. Sta., East Lansing, MI.

Manrique, L. A., C. A. Jones and P. T. Dyke. 1991. Predicting Cation-Exchange Capacity from Soil Physical and Chemical Properties. *Soil Sci. Soc. Am. J.* 55:787–794.

Niftal Project. 1989. *Dynamics of Soil Organic Matter in Tropical Ecosystems*. Univ. Hawaii Press, Honolulu.

Sanchez, P. 1976. *Properties and Management of Soils in the Tropics*. John Wiley & Sons, New York.

Sanchez, P. A., J. H. Villachica and D. E. Bandy. 1983. Soil Fertility Dynamics after Clearing a Tropical Rainforest in Peru. *Soil. Sci. Soc. Am. J.* 47:1171–1178.

Smith, R. F. and B. L. Kay. 1986. Revegetation of Serpentine Soils: Difficult but not Impossible. *California Agriculture*. 40:18–19.

Soil Survey Staff. 1992. *Keys to Soil Taxonomy*. 5th ed. Soil Management Support Services (AID) Tech. Monograph No. 19, USDA. Pocahontas Press, Blacksburg, VA.

Sposito, G. 1989. *The Chemistry of Soils*. Oxford Press, New York.

St. Arnaud, R. J. and G. A. Septon. 1972. Contribution of Clay and Organic Matter to Cation-Exchange Capacity of Chernozemic Soils. *Can. J. Soil Sci.* 52:124–126.

Stumpe, J. M. and P. L. G. Vlek. 1991. Acidification Induced by Different Nitrogen Sources in Columns of Selected Tropical Soils. *Soil Sci. Soc. Am. J.* 55:145–151.

Tan, K. H. 1993. *Principles of Soil Chemistry*. 2nd ed. Marcel Dekker, New York.

Tessens, E. and S. Sauyah. 1982. Positive Permanent Charge in Oxisols. *Soil Sci. Soc. Am. J.* 46:1103–1106.

Thomas, G. W. and W. L. Hargrove. 1984. The Chemistry of Soil Acidity. *Soil Acidity and Liming. Agronomy* 12:1–56. Am. Soc. Agron., Madison, WI.

Thompson, M. L., H. Zhang, M. Kazemi and J. A. Sandor. 1989. Contribution of Organic Matter to Cation Exchange Capacity and Specific Surface Area of Fractionated Soil Materials. *Soil Sci.* 148:250–257.

Wildman, W. E., M. L. Jackson and L. D. Whittig. 1968. Iron Rich Montmorillonite Formation in Soils Derived from Serpentinite. *Soil Sci. Soc. Am. Proc.* 32:787–794.

Wolcott, A. R., H. D. Foth, J. F. Davis and J. C. Schickluna. 1965. Nitrogen Carriers: I. Soil Effects. *Soil Sci. Soc. Am. Proc.* 29:405–410.

CHAPTER 4

Soil Reaction

Soil reaction is the degree of soil acidity or alkalinity which is caused by a particular chemical, mineralogical, and/or biological environment. Soil reaction affects element availability and toxicity, microbial activity, and root growth. It is one of the most important properties affecting soil fertility and is commonly managed to increase plant growth. Iron is frequently deficient for plant growth on alkaline soils and iron deficiency symptoms on plants are commonplace throughout the world because about 30% of the world's soils are calcareous and alkaline.

Soil reaction also affects the pH and ionic content of groundwater. When soils are very acid, they may develop sufficient Al in the rooting zone to be toxic to plants and occasionally contribute a toxic amount of Al to shallow water supplies. The acidity produced from high rates of N ammonical fertilizers on acid soils may threaten soil sustainability.

Soil reaction is usually expressed as a pH value. Descriptive terms commonly associated with certain ranges in pH are extremely acid, <4.5; very strongly acid, 4.5-5.0; strongly acid, 5.1-5.5; moderately acid, 5.6-6.0; slightly acid, 6.1-6.5; neutral, 6.6-7.3; slightly alkaline, 7.4-7.8; moderately alkaline, 7.9-8.4; strongly alkaline, 8.5-9.0; and very strongly alkaline, >9.1.

4.1 DETERMINATION OF SOIL pH

The pH is an expression of the H^+ activity of a solution. Hydrogen ions in solution hydrate to form the hydronium ion, H_3O^+, but hydration water is usually not shown for simplicity. The pH soil test is an empirical method which typically determines the pH of a suspension or a paste.

4.1.1 Common Methods

Most soil tests measure the pH of a standardized soil suspension. The pH of mineral soils is commonly determined in the U.S. by mixing soil and water together in a 1:1 weight to volume ratio. For example,10 g of air-dry soil that has passed a 2 mm screen are placed in a 50-ml beaker, and 10 ml of distilled water is added. The soil and water are stirred occasionally for an hour. Then the pH of the slurry is determined with a glass electrode and pH meter while the slurry is stirred.

Standard conditions for pH determination are given in the format as soil weight to solution volume and the solution used. For example, 1:1 in water means equal weights of soil and distilled water are equilibrated. The amount of solution relative to soil is increased for determining the pH of organic soils; commonly, a soil:water ratio of 1:2 to 1:5 accommodates the high absorbency of the organic matter.

The routine analysis of sodic soils generally includes the preparation of a saturated paste using water and the measurement of pH is made in the paste. This method is commonly used in southwestern U.S.

There are inexpensive kits for determining soil pH. These are sufficiently accurate to diagnose many plant growth problems in the field that are associated with soil reaction.

4.1.2 Factors Affecting Soil pH Measurement

Two important factors that affect the pH value of soils are soil:liquid ratio and the presence of salts. The more water added to a given amount of soil decreases the soil-water ratio resulting in greater dilution of the H^+ and increases in the pH value determined for noncalcareous soils. In soils containing soluble carbonate minerals, the pH value increases as water is added because of the dissolution of carbonates.

The soluble salt content of soil can affect the pH value and this may be overcome by using salt solutions instead of distilled water. Two solutions commonly used are 0.01 M $CaCl_2$ and 1.0 M KCl. The cations from salt solutions exchange with and put exchangeable acidic cations into solution. When salt solutions in the range of 0.07 to 1 M are used, soil pH values of acid soils may be 0.5 to 1.5 units lower than when they are measured with distilled water. The pH values of the horizons of a well-drained Spodosol, determined with 0.01 M KCl and distilled water are shown in Figure 4.1. The greatest difference in pH between the use of water and KCl occurred for the horizons in the upper part of the profile containing the most ECEC (and where there was the greatest displacement of acidic cations by the KCl).

Acid soils with a net positive charge, however, will have a higher pH using KCl due to the replacement of OH^- by chloride ions.

When the pH values of various soils are compared, it is important to compare values that were determined by the same method. Further, the pH of an air-dry soil sample in the laboratory will be different from the pH that exists in the same soil in the field during the growing season. There will be differences in water and salt content, and the respiration of roots and microorganisms produces carbonic acid.

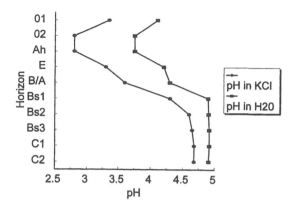

Figure 4.1 Mean pH of the horizons of a Spodosol determined with KCl and H_2O. (Redrawn from Olsson and Melkerud, 1989. Used by permission of Elsevier Science Publishers B.V.)

4.2 THE SOIL pH CONTINUUM

The changes in the mineralogical and chemical properties of soils that occur with time, as discussed in Chapter 2, are related to and parallel changes in soil pH. In general, soil pH decreases as soils become more weathered and leached and this creates a soil pH continuum ranging from about 10 to 4 for most natural soils. The actual pH of a particular soil is mainly a function of soil composition and the ion exchange and hydrolysis reactions associated with various soil components.

4.2.1 Alkaline Soils

The dissolution of soluble carbonates is the major cause of soil alkalinity. Calcareous soils contain free calcium carbonate and its dissolution produces OH⁻ as follows.

$$CaCO_3 + H_2O = Ca^{2+} + HCO_3^- + OH^- \tag{4.1}$$

The pH ranges from slightly more than 7 to about 8.5 and is related to the solubility of the $CaCO_3$. The soils are slightly or moderately alkaline. Lesser amounts of $MgCO_3$ also are present in calcareous soils. The soils are 100% saturated with basic cations and Ca and Mg are the dominant exchangeable cations.

Sodium carbonate is more soluble than $CaCO_3$ and greater dissolution results in more OH⁻ activity producing a soil pH in the range 8.5 to about 10. When the presence of Na_2CO_3 produces 15% or more XNa saturation, the soil is considered sodic and pH becomes >8.5.

$$soil\text{-}Na + H_2O = soil\text{-}H + Na^+ + OH^- \qquad (4.2)$$

Sodic soils have a sodium adsorption ratio (SAR) of thirteen or more, which is equated to a Na saturation of 15% or more, and there also is deflocculation and dispersion of soil colloids. The soils are 100% basic cation saturated and Na is frequently the most abundant exchangeable cation. Sodium affected soils are strongly or very strongly alkaline.

Plant growth on sodic soils is affected by low solubility and availability of some nutrients and poor soil physical conditions. Conversion of sodic soil to nonsodic soils may be accomplished with the application of gypsum. The calcium sulfate from gypsum, with controlled leaching, results in the replacement of XNa by Ca and removal of replaced XNa as Na_2SO_4. The treatment lowers the Na saturation percentage and increases the Ca saturation percentage. The soil becomes nonsodic when the exchangeable Na saturation decreases to less than 15%.

The dissolution of CO_2 released by respiring roots and microorganisms produces carbonic acid. Such acid production may be sufficient to cause local pockets of acidity in alkaline soils, but not enough to produce a thoroughly acid soil. Much of the carbonic acid decomposes to water and CO_2, and the carbon dioxide eventually escapes from the soil.

Precipitation is naturally acidic (pH 5.67 for pure water in equilibrium with CO_2 in the atmosphere) and when in equilibrium with the normal CO_2 content of the atmosphere, natural precipitation is a continuous source of acidity for soils. In addition, nitrification, S oxidation, and other processes continually produce H^+. These additions of acidity are neutralized in alkaline soils, which is shown for a calcareous soil in the following example:

$$CaCO_3 + 2H^+ = Ca^{2+} + H_2O + CO_2 \qquad (4.3)$$

The H^+ is removed by the formation of water. Many alkaline soils have a very large acid-neutralizing capacity, depending on their carbonate content.

Many alkaline soils occur in arid regions, owing to low precipitation and biological activity. Soils in humid regions are generally acid. Alkaline soils occur in humid regions where young soils are forming from alkaline parent materials, where soil erosion exposes underlying alkaline materials, and where soils periodically receive alkaline sediments, as in the case of many alluvial soils.

4.2.2 Neutral Soils

The formation and maintenance of alkaline soils occur when Ca^{2+} and Mg^{2+} released from weathering form carbonates in the absence of effective leaching. Alkaline soils become neutral when the carbonates are removed by leaching. The pH is in the range 6.6 to 7.3. The effective cation exchange capacity (ECEC) is 100% basic cation saturated and overwhelmingly dominated by XCa and XMg.

Neutral soluble salts, if present, contribute to neutrality. A saline soil contains sufficient soluble salt to affect plant growth. Saline soils develop where soluble salts accumulate from natural or from human activities, such as irrigation or fertilizer application.

4.2.3 Acid Soils

Neutral soils remain neutral when the inputs of acidity from the precipitation, associated biological activity, and other factors are balanced by the neutralizing inputs of mineral weathering and the return of basic cations through nutrient cycling or additions through dust or volcanic ejecta. Neutral soils become acid when the acidic inputs in time exceed the neutralizing inputs. As a consequence, acid soils tend to occur in humid regions where the acidic inputs from precipitation and associated biological activity are high and where leaching is effective in removing the soils' neutralizing components. The acids that are constantly added to neutral soils in humid regions are a source of H^+, which gradually replace exchangeable basic cations. Subsequently, the basic cations are leached out of the soil with accompanying anions in the surplus water. The following equation illustrates the leaching of $Ca(NO_3)_2$:

$$CaX + 2HNO_3 = 2HX + Ca(NO_3)_2 \qquad (4.4)$$

The loss of basic cations is associated with decreasing soil pH.

Acidity, on the other hand, stimulates the weathering of primary minerals and generates alkalinity. The abrasion pH (pH of ground mineral in water) of albite is 9 or 10. The abrasion pH of most of the primary minerals of the sand and silt fraction is between 7 and 11. Their weathering supplies the soil with alkali and alkaline earth cations (Na, K, Ca, Mg) that contribute to the maintenance of neutrality and/or alkalinity. The weathering of albite feldspar illustrates the release of Na^+.

$$NaAlSi_3O_8 + 4H^+ + 4H_2O = Na^+ + Al^{3+} + 3Si(OH)_4 \qquad (4.5)$$

The weathering consumes H^+ and produces OH^-, resulting in reduced acidity. Some of the Na, Ca, Mg, and K ions released in weathering are adsorbed as exchangeable cations and contribute to the supply of basic cations.

Note in Equation 4.5 that some Al is released by mineral weathering. As neutral soils become acid, some of the Al^{3+} forms hydroxy-Al, which is strongly adsorbed by clays and soil organic matter (SOM). The adsorption reduces the ECEC. Thus, in the early stages of acidification, there is (1) a decrease in the ECEC, and (2) a decrease in the amount of basic cations via leaching, even though the ECEC at this point is 100% basic cation saturated.

The ECEC decreases with acidification and this is associated with increasing mobilization of Al. Near pH 7, relatively inert $Al(OH)_3^0$ exists and, with increasing

H^+ activity, protonates to form $Al(OH)_2^+$ which protonates to form $Al(OH)^{2+}$. These mobilized forms of hydroxy-Al, in their simplest forms, are adsorbed first as $Al(OH)_2^+$, and then as $Al(OH)^{2+}$, as soil acidification progresses. Polymerization of hydroxy-Al produces large cations of variable charge. These hydroxy-Al cations are strongly adsorbed by both the mineral and organic fractions, and account for the decrease in the ECEC.

There is an absence of XAl at this stage of soil acidification. As soil pH decreases more, $Al(OH)^{2+}$ protonates and forms trivalent Al, Al^{3+}, which begins to effectively dominate the ECEC. The sequential protonation reactions that occur with increasing soil acidification are as follows:

$$Al(OH)_3^0 + H^+ = Al(OH)_2^+ + H_2O \qquad (4.6)$$

$$Al(OH)_2^+ + H^+ = Al(OH)^{2+} + H_2O \qquad (4.7)$$

$$Al(OH)^{2+} + H^+ = Al^{3+} + H_2O \qquad (4.8)$$

The effect of pH on the dominant forms of Al in solution and their average charge is shown in Figure 4.2.

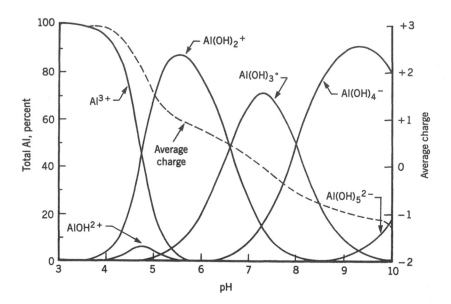

Figure 4.2 Relative distribution of the forms of soluble aluminum as a function of pH. (From Marion et al., 1976. Used by permission of Williams & Wilkins Company.)

During soil acidification, protonation increases the mobilization of Al and the Al forms serve as a sink for the accumulation of H^+. This accumulated acidity must be neutralized to increase soil pH, as in the case of liming.

Below 5.5, there is increasing saturation of the ECEC with Al^{3+}. At pH 4, there is about 100% saturation. Thus, in the pH range 5.5 to 4 there is decreasing absolute and relative amounts of basic cations and an increasing dominance of acidic cations, especially XAl.

The major source of H^+ for the soil solution when soil pH decreases below 5.5 is the hydrolysis of XAl, Al^{3+}, and Al polymers.

$$Al^{3+} + H_2O = Al(OH)^{2+} + H^+ \tag{4.9}$$

As soil pH increases above 5.5, the sequential hydrolysis of hydroxy-Al becomes the major source of H^+ for the soil solution.

$$Al(OH)^{2+} + H_2O = Al(OH)_2^+ + H^+ \tag{4.10}$$

$$Al(OH)_2^+ + H_2O = Al(OH)_3^0 + H^+ \tag{4.11}$$

Pyrite (FeS_2) is a component of some mine spoils and some alluvial soils. Upon oxidation of the sulfate, free sulfuric acid is produced and the soils may be called acid sulfate soils. Soil pH of 2 or less can be produced and conditions are very unsatisfactory for plant growth due to high soluble Al, Fe, and Mn, along with a deficiency of basic cations. The acidity *per se* becomes detrimental to biological activities at pH less than 4 or 3.5.

4.2.4 Summary Statement

1. Below 3.5 to as low as 2, soil pH is controlled by the presence of strong acid.
2. In the pH range 4 to about 7, the hydrolysis of exchangeable and hydroxy forms of Al and humus carboxyl groups are the major sources of H^+. Increasing soil pH is associated with a decrease of acidic exchangeable cations and an increase of basic exchangeable cations.
3. Neutral soils are 100% saturated with exchangeable basic cations and Al hydrolysis is unimportant.
4. In range 7 to 8.3 (pH range for calcareous soils), the pH is controlled by hydrolysis of $CaCO_3$.
5. In the range 8.5 to 10 (pH range for sodic soils), pH is controlled by hydrolysis of Na_2CO_3 and/or XNa.

4.3 FORMS OF SOIL ACIDITY

Generally, the total acidity is equated to the sum of the exchangeable and the titratable acidity. The exchangeable acidity is removed by leaching the soil with an unbuffered salt solution, such as KCl. Thus, it is called KCl or salt extractable acidity. The exchangeable acidity consists mainly of the exchangeable acidic cations plus a small and variable amount of the hydroxy-Al that is extractable. In essence, exchangeable acidity is equivalent to exchangeable Al in many soils.

After the salt extractable or exchangeable acidity has been removed, the titratable or nonexchangeable acidity is neutralized by titration of the soil to a particular pH value. Generally, titratable acidity is neutralized with $BaCl_2$-triethanolamine (TEA) buffered at pH 8.2.

In temperate zone soils, titratable acidity is usually many times larger than exchangeable acidity and must be neutralized to increase soil pH. The accumulation of extractable acidity, as represented by a decrease in basic cations, and the accumulation of titratable acidity with decreasing pH, caused by application of N fertilizers to an Alfisol, are shown in Figure 4.3.

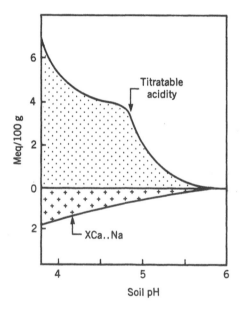

Figure 4.3 Titratable acidity and decrease of XCa..Na (exchangeable bases) as a function of pH in the Ap horizon of an Alfisol. (Data from Wolcott et al., 1965.)

4.4 REACTION OF SOIL ORDERS

The mean pH of a large number of A, B, and C horizons of nine orders is given in Table 4.1. The mean pH ranges from 8.1 for Aridisols to 5.0 for Ultisols.

In general, the least weathered and leached soils are neutral or alkaline and the most weathered and leached soils are strongly acid. The reaction profile can be used to classify soils into three groups: (1) soils with alkaline and/or neutral horizons throughout, (2) soils with acid surface horizons underlain with alkaline horizons, and (3) soils with acid horizons throughout. These three kinds of soils represent the sequential development of reaction profiles during soil evolution

Table 4.1 Mean pH of Horizons of Soil Orders

Order	pH	Observations*
Aridisols	8.1	4,421
Vertisols	7.5	1,279
Mollisols	7.0	13,927
Entisols	6.9	3,747
Inceptisols	6.0	5,506
Alfisols	6.0	11,465
Oxisols	5.5	300
Spodosols	5.1	2,074
Ultosols	5.0	5,565

*A, B, and C horizons, 1:1 soil to water ratio.
Source: Manrique, Jones, and Dyke, 1991.

from alkaline parent materials in humid regions. This classification recognizes the importance of soil reaction within the root zone for plant growth and the role of soil reaction of horizons below the root zone in environmental quality.

4.4.1 Alkaline and Neutral Soils

The average pH of Aridisols, Vertisols, and Mollisols is 8.1, 7.5, and 7.0, respectively. The entire root zone of Aridisols is typically alkaline or neutral. The same is generally true for Vertisols and Mollisols with ustic soil moisture regimes. In soils with aridic and ustic soil moisture regimes, limited leaching may produce calcium carbonate accumulation zones or k layers, such as Bk or Ck horizons. Such horizons are slightly or moderately alkaline and have a very high acid-neutralizing capacity. Surface runoff water is alkalized by contact with soils. The lack of surplus water means that the soils have little impact on groundwater. However, where excess water accumulates in areas of limited precipitation, and surplus water leaves the soil, the water is alkaline.

4.4.2 Acid Soils Underlain with Alkaline Layers

The leaching of basic cations and development of acidity occur progressively from the soil surface downward. This causes Vertisols and Mollisols with udic soil moisture regimes to typically have some acid horizons overlying neutral and/or alkaline horizons.

The average pH of Entisols is 6.9 and for Inceptisols and Alfisols is 6.0 (see Table 4.1). Entisols and Inceptisols occur in all climatic zones and include soils that range from being alkaline throughout to being strongly acid throughout. Many of these soils in humid-temperate regions, however, are only moderately weathered and leached and have acid upper layers that are frequently underlain by alkaline layers with high acid neutralizing capacity. Usually, the increased supply of basic cations and reduced acidity in underlying alkaline horizons have little affect on plant growth. Deep-rooted perennials, like alfalfa, may benefit sometimes from underlying alkaline layers. These soils have high capacity to neutralize acids and precipitate soluble Al in surplus water before the water moves to the water table.

4.4.3   Acid Soils without Underlying Alkaline Layers

The average pH is 6.0 for Inceptisols and Alfisols (see Table 4.1). Some of these soils are acid to great depths, especially where they occur in the humid tropics. The same is true for Oxisols, Spodosols, and Ultisols that have a mean pH ranging from 5.0 to 5.5. The Oxisols and Ultisols are the most intensively weathered soils and have the lowest mean pH, except for Spodosols. The entire root zone of these soils is usually strongly acid, similar to the situation shown in Figure 4.1. Roots do not encounter horizons with high basic cation saturation and sometimes root growth in subsoils is restricted due to Al toxicity and a deficiency of Ca and/or Mg. Acids in surplus water leaving the soil are unlikely to be neutralized and surplus water commonly contributes acidity and soluble Al to groundwater and lakes (Figure 4.4).

Figure 4.4 Excess acid water leaving an acid soil, resulting in acidified groundwater.

The pH of groundwater and lakes is related to the reaction of the soils and underlying materials in the surrounding area. Acid rain can be a problem in humid regions where soils and underlying materials do not have the capacity to neutralize the acids in surplus water that moves to the water table. In many cases, acid surplus water high in acidic cations and low in basic cations is neutralized on contact with calcareous soil or parent material layers. This results in alkaline groundwater and lake water high in basic cations and very low in Al^{3+}. Since precipitation is naturally acidic, alkaline lakes owe their existence in large part to neutralization and alkalization of precipitation by contact with alkaline soils and other geologic materials.

The pH of the horizons of Histosols (organic soils) varies from being strongly acid to being strongly alkaline. Histosols that develop in closed basins are greatly affected by the pH and the solutes in the water where the soil develops. The formation of soils from the accumulation of sphagnum moss in acid water produces highly acid organic layers. On the other hand, some Histosols have been affected by alkaline waters and contain layers of marl that are strongly alkaline.

In summary, root zones tend to be entirely neutral and/or alkaline, entirely acid, or acid in the upper part and alkaline in the lower part. This affects the root environment, the availability and/or toxicity of elements, and the pH and soluble Al, Mn, etc. content of the surplus water leaving the soil.

4.5 ATMOSPHERIC INPUTS

Acid rain and dust are two atmospheric inputs that affect soil reaction. Acid rain is important in humid regions and contributes to soil acidity. Dust is pervasive in arid and nearby regions and contributes to soil alkalinity and soil acidity depending upon the source of the dust particles. A third atmospheric input is volcanic ash and ejecta, which are especially important near active volcanoes.

4.5.1 Acid Rain

Unpolluted rain in equilibrium with the normal CO_2 content of the atmosphere has a pH of 5.6 (5.67 for pure water). A meter of this rain, which is about a year's supply in humid regions, can dissolve about 320 to 400 pounds/acre of $CaCO_3$ (400 to 500 kg/ha) once it reacts with CO_2 in the soil. This is roughly equal to the amount of lime needed to maintain the pH of agricultural soils in Illinois. The burning of fossil fuel releases oxides of S and N that eventually return in the precipitation; they cause the pH of the precipitation to be less than 5.6, and to become an acid precipitation. In northeastern U.S., the precipitation in 1981 contained an acid equivalent, from the effects of acid rain, of 28.8 pounds/acre of $CaCO_3$ (32.3 kg/ha). At Knoxville, Tennessee, the annual figure has been reported to be 83 pounds/acre (93 kg/ha). By comparison, the acid producing potential of the average N fertilizer application in U.S. is equal to 179 pounds/acre (201 kg/ha). The fixation of N by alfalfa can amount to the equivalent of 500 pounds/acre, or more of $CaCO_3$ (560 kg/ha). Since only part of the total precipitation is due to acid rain component, it can be seen that acid rain contributes a small amount to the total acid input of soils in the humid region of the U.S. which becomes relatively unimportant if soil pH is being managed by liming.

4.5.2 Dust

Some Aridisols have features, including argillic horizons, that are believed to have developed under a humid climate. The climate changed from humid to arid and dust played an important role in resaturation of the ECEC with basic cations and conversion from acid to alkaline soil.

The seasonal Harmattan winds cause the deposition of about 500 to 1,000 kg ha^{-1} of dust annually in northern Nigeria. The dust is 25% clay, 57% silt, and 18% fine sand. Compared to the soils at six locations, the dust is richer in basic cations, lower in exchangeable acidity, and contains more extractable P. The dust also has a greater amount of constant charge clay as compared to variable charge clay. It is hypothesized that Alfisols of the region would have evolved into Ultisols in the absence of the long-time influx of Harmattan dust.

4.5.3 Volcanic Ejecta

Soils near active volcanoes typically develop thick layers of pyroclastic materials, giving rise to andic properties and Andisols. Farther from volcanoes, ash contributes fresh unweathered material to soils in all climatic zones. The loess of the northwestern U.S. and of Argentina has a significant ash component; the loess in the north-central U.S. does not.

As with weathering of primary minerals generally in soils, weathering of volcanic ash contributes to soil alkalinity. After the 1980 eruption of Mount. St. Helens in Washington, tephra was applied in 5- and 15-cm thick layers to Spodosols that had not been affected by the eruption. Studies of the composition of the soil solution of tephra-treated soils, compared to nontreated soils, showed that leaching of the tephra-transported basic cations replaced acidic cations in the upper layers of underlying Spodosols.

4.6 EFFECTS OF SOIL pH ON PLANT GROWTH

Within the pH range of 4 to 10, the primary effect of soil pH on plant growth is not the H$^+$ or OH$^-$ activities *per se* but the associated chemical environments. In general, the major influence of pH is on ion activities that affect the toxicity of elements like Al and Mn, or nutrient availability as shown for mineral soils in Figure 4.5.

Nitrogen availability is greatest between 6 and 8 because the mineralization of N is maximum in this range. The availability of P in acid soil is reduced by precipitation and adsorption by Fe and Al. Calcium carbonate in calcareous soils reduces the solubility of calcium phosphorus compounds by increasing the activity of Ca^{2+} or by adsorption on CaCO$_3$ surfaces. These examples specifically show how pH affects nutrient availability and more complete discussions for the various nutrients appear in later chapters. Here, the goal is to focus on the specific causes of acid and alkaline soil infertility.

4.6.1 Acid Soil Infertility

Aluminum toxicity for plants is a common problem on acid soils and is related to the ECEC and Al saturation. When Al saturation exceeds 60% in soils with a significant ECEC, there is likely to be 1 ppm or more of Al^{3+} in solution and

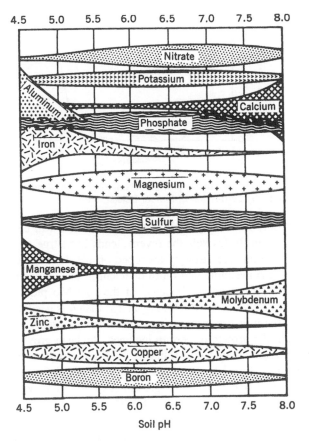

Figure 4.5 The relation between soil pH and the relative availability of plant nutrients in mineral soils. The wider the bar, the greater the availability. (From *Soil Handbook*, University of Kentucky, 1970.)

many crop plants will suffer from Al toxicity. The forms and amounts of Al in solution as a function of solution pH are shown in Figure 4.6. The hydroxy-Al forms in acid soils are in low concentration and play a minor role in terms of Al toxicity, as compared to Al^{3+}.

Symptoms of Al toxicity include short and stubby roots. High-soluble Al inhibits the uptake and translocation of P in plants and may cause P deficiency. Soluble Al also reacts with P to reduce P availability. Thus, it is difficult to know in many cases whether poor plant growth in very acid soils is due to Al toxicity *per se* or if P deficiency is due to the presence of high-soluble Al.

Below pH 5, Al^{3+} becomes the dominant form of Al in solution (see Figure 4.6). The forms of Al in solution when pH is above 5 are less toxic, and at pH 7 the dominant form of Al in solution is $Al(OH)_3^0$. Above pH 7, the Al in solution is negatively charged and repelled by the soil's negative charge.

Dissolution of manganese minerals at low pH may result in Mn toxicity, especially in soils high in manganese minerals. In strongly acid Ultisols, plant

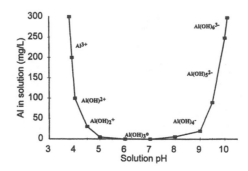

Figure 4.6 Effect of pH on the solubility of aluminum in solution. (Redrawn from McLean, 1976. Used by permission of Marcel Dekker, Inc.)

growth is commonly limited by toxic levels of Al and/or Mn before Ca and/or Mg deficiencies occur. In Oxisols the reverse tends to be true.

Molybdenum anions are strongly adsorbed on pH-dependent clays and Mo tends to become less available and more deficient in acid soils. The availability of Mo is important for legumes because of the role Mo plays in N fixation. Although it is generally considered that a pH less than 4 is needed to affect plant growth directly, the infection process of some *Rhizobia* strains is acid sensitive. Acidity may limit the growth of certain legumes. Soybean growth on some acid Ultisols in northern Alabama was found to be limited most frequently by Mo deficiency; secondly, by Al toxicity; and thirdly, by Mg deficiency. A summary of the toxic factors found to limit the growth of alfalfa in some acid soils from Arkansas and Georgia is given in Table 4.2. The Bladen and Cecil are Ultisols, the Leon is a Spodosol, and the Loring and Zanesville are Alfisols. Plants show great diversity in tolerance and resistance to the factors so that the results shown in Table 4.2 for alfalfa would be different for a crop like cowpeas which is more tolerant of high levels of soluble Mn.

Table 4.2 Toxic Factors Limiting Growth of Alfalfa on Acid Surface Soils

Soil	pH	Region	State	Factor
Bladen	4.8	Coastal plain	Georgia	Al toxicity
Cecil	5.2	Piedmont	Georgia	Mn toxicity
Leon	4.2	Coastal plain	Georgia	Ca deficiency
Loring	5.3	Loess hills	Arkansas	Mn toxicity
Zanesville	4.9	Ozarks	Arkansas	Mn toxicity

Source: From Foy and Burns, 1964.

Studies have shown that for a given pH, acid soils have less XAl if they contain more organic matter because of the complexing or chelation of the Al by SOM. Many acid surface soils allow for good root development in cases where subsoils with the same pH results in poor root growth due to Al toxicity. This has been confirmed by the addition of organic matter to acid mineral soil mixes and noting the reduction in XAl and observing that plants can grow well in soil with low pH if well supplied with organic matter as shown in Figure 4.7.

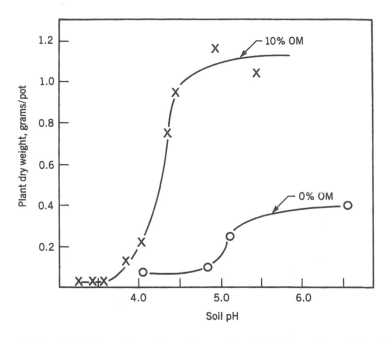

Figure 4.7 The effect of organic matter on the growth of plants in soil consisting of sand and Al-montmorillonite mixtures. (From Hargrove and Thomas, 1981. Used by permission of the Soil Science Society of America.)

In acid Histosols, plants are unlikely to suffer from Al toxicity due to the high content of organic matter and complexing of Al. Histosols also have a low content of mineral matter. These factors contribute to the availability of plant nutrients as affected by pH in Histosols as shown in Figure 4.8. Maximum overall nutrient availability is shifted downward to a pH of about 5.5 as compared to 6.5 for mineral soils.

4.6.2 Alkaline Soil Infertility

Alkaline soil infertility tends to be associated with calcareous soils where both pH and presence of $CaCO_3$ affect nutrient availability. Iron is the most common nutrient element deficiency of plants growing on alkaline soil and higher plants have devised two "strategies" to increase the availability of Fe in soils. The grasses excrete siderophores which can solubilize Fe. The other monocots and dicots can decrease the pH in the rhizosphere by proton excretion. The solubilized ferric ions resulting from proton excretion and the ferric chelates caused by siderophore excretion can be reduced in the plasma membrane.

Calcareous soils are frequently productive for a wide range of crops including cotton, corn, sugar beet, potato, and many legumes. On the other hand, many plants including roses, pin oak, grapes, sorghum, and maple are susceptible to Fe and/or Mn deficiencies. Zinc and P may also have low availability at high pH.

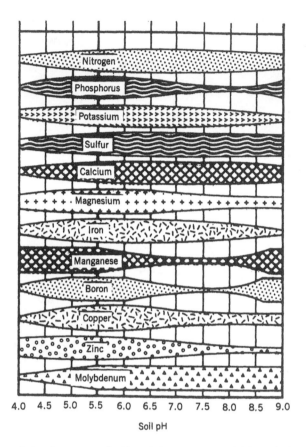

Figure 4.8 The relation between soil pH and the relative availability of plant nutrients in organic soils or Histosols. The wider the bar, the greater the availability. (Data from Lucas, 1982.)

Some alkaline soils with soluble salts may contain enough soluble B to produce B toxicity. The low availability of P in calcareous soils is discussed in Chapter 7.

4.6.3 Soil pH and Plant Growth

As we have noted, the pH of most soils is in the range of 4 to 10 and depends on the soil parent material and the changes that have occurred during soil genesis. As a result, the plants that have evolved in desert regions have adapted to thrive on neutral and alkaline soils, and many desert plants have great tolerance for soluble salts and soluble Na (halophytes). Plants that are native to the humid tropics, like cassava, have tolerance for soluble Al. Some temperate-region plants that do well on highly acid soils, such as rhododendron, appear to be Al accumulators.

The Al tolerance of some important tropical crops is given in Table 4.3. Low tolerant plants can tolerate up to 40% Al saturation, and moderate tolerance from

40 to 70% Al saturation. High Al-tolerant crops can grow well when the Al saturation is between 70 and 100%. Aluminum tolerance appears to be related to an exclusion mechanism for Al uptake, entrapment of Al in roots, or the accumulation of Al in the tops.

Table 4.3 Aluminum Tolerance of Crops Grown in the Tropics

Crop	Aluminum tolerance
Maize	Low
Groundnut	Moderate
Cassava	High
Cowpeas	Moderate to high
Bean	Low to moderate
Trees	
Leucaena	Low
Cocoa	Low
Oil palm	Moderate to high

Source: Caudle, 1991.

Other physiological responses to plant growth on acid soils include the ability to increase the pH of the rhizosphere, preference for NH_4^+ over NO_3^-, the ability to absorb phosphate from solution very low in phosphate, and a low Ca requirement. High genetic variation exists between species and cultivars, and this is used to develop new varieties that can better cope with stressful soil conditions related to soil pH.

4.6.4 Plant Response to Soil pH

There are many kinds of soil and many different plant responses to soil conditions. In general, as soil pH is increased from 4, the root and microbial environment change. There is a tendency for plant growth to increase with an increase in soil pH as the factors that limit growth are ameliorated. A plateau is reached where further increases in soil pH have little if any effect on plant growth. The juncture between the rapidly increasing plant growth part of the curve and the plateau region is the critical pH. Different soils may have different critical pH values for a given crop because the factors limiting growth are different in different soils.

4.7 EXTREME WEATHERING EFFECTS ON pH

Two acid soils can have the same pH and have greatly different soil fertility parameters because one soil has charge that is dominated by high constant charge minerals and the other soil has charge that is dominated by low and variable charge minerals. Thus, there is a need to elaborate on the effects of extreme weathering on soil pH, the ECEC, and the exchangeable cation suite as they relate to the fertility of soils.

The difference in clay activity of minimally and moderately weathered soils and the more weathered soils is used to differentiate soils into high and low activity clay soils. Oxisols, generally, have low activity clay that is recognized by having an ECEC of 12 or less $cmol_c$ kg^{-1} or a CEC, at pH of 7, of 16 or less $cmol_c$ kg^{-1}. The clays are mainly kaolinite and Fe and Al oxides. Soils with clay meeting these criteria are called low activity clay (LAC) soils. Generally, Ultisols and Oxisols are considered LAC soils even though the clay of many Ultisols has a CEC above the limits.

In the most weathered Oxisols, weathering has virtually ceased. Some part of the oxic horizon contains 1.5 or less cmol kg^{-1} of exchangeable cations (including Al) per kg of clay that can be removed by leaching with unbuffered KCl. This extreme degree of weathering is indicated with the prefix, acr. An Acrudox is a Udox that is extremely weathered. The A horizons of natural Acrudox have enough organic matter to produce a net negative charge. The low organic matter content of the lower part of the oxic horizon usually results in a net positive charge. The lower part of the oxic horizon has higher pH when determined with KCl than with H_2O. Such a soil is an Anionic Acrudox, the most weathered kind of soil. When soils become extremely weathered, the pH is higher than less weathered soils with greater ECEC and XAl. Such soils, however, have a very small amount of exchangeable basic cations and they may have an insufficient quantity of XAl to produce Al toxicity. Calcium and Mg are likely to be deficient for cropping.

Based on degree of weathering, mineral soils can be classified into (1) high activity clay soils with high ECEC and high basic cations, (2) LAC soils with low ECEC and low basic cations, and (3) LAC soils with very low ECEC and very low basic cations. Thus, three different soils can have similar pH and have greatly different fertility parameters and need different soil management, including liming practices (see Figure 4.9). Although world agriculture is concentrated on high activity clay soils of the temperate regions, many of the LAC soils of the tropical and subtropical regions are very productive. Examples include the Ultisols of the coastal plain of the southeastern U.S. for production of cotton and soybeans, Oxisols in Hawaii for sugar cane and pineapples and in Brazil for soybeans, wheat, sugar cane and a wide variety of other crops (see Figure 4.10).

These generalizations apply to soils developed in parent material of mixed mineralogy because some young soils inherit low activity clays. Some Oxisols, Torrox and Ustox, have a fairly good supply of basic cations. Some Oxisols may develop without having been Ultisols. Even so, the mineral weathering sequence concept is well established and is useful in relating differences in soil pH and cation exchange relationships between several of the soil orders.

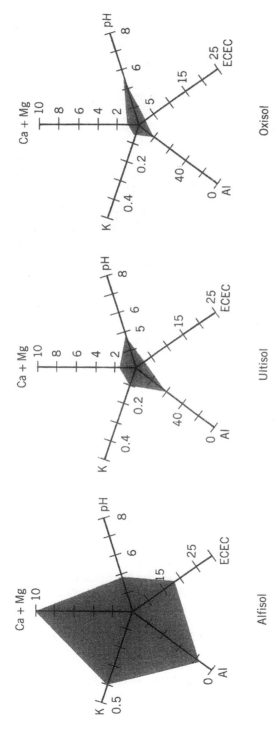

Figure 4.9 Soil fertility parameters, including exchangeable potassium, exchangeable calcium plus magnesium, and effective cation exchange capacity in milliequivalents per 100 grams; the pH; and the percentage of aluminum saturation of A horizons of an Ultic Paleudalf, a Plinthic Paleudult, and an Acrorthox from left to right, respectively. Each soil represents a highly weathered soil for the orders Alfisol, Ultisol, and Oxisol, respectively.

Figure 4.10 High-yielding soybeans grown on Oxisols in the Brazilian highlands.

REFERENCES

Adams, F. 1984. Soil Acidity and Liming. 2nd ed. *Agronomy 12.* Am. Soc. Agron., Madison, WI.

Alvim, P. de T. 1979. Pasture Production in Acid Soils of the Tropics. In *Agricultural Production Potential of the Amazon Region.* P. A. Sanchez, Ed. Centro Internacional de Tropical (CIAT), Cali, Colombia, S.A.

Barber, S. A. 1984. *Soil Nutrient Bioavailability.* John Wiley & Sons, New York.

Bienfait, H. F. 1988. Mechanisms in Fe-Efficiency Reactions of Higher Plants. *J. Pl. Nut.* 11:605–629.

Bohn, H. L., B. L. McNeal and G. A. O'Connor. 1985. *Soil Chemistry.* 2nd ed. John Wiley & Sons, New York.

Burmester, C. H., J. F. Adams and J. W. Odom. 1988. Response of Soybean to Lime and Molybdenum on Ultisols in Northern Alabama. *Soil Sci. Soc. Am. J.* 52:1391–1394.

Caudle, Neil. 1991. *Groundworks 1, Managing Soil Acidity.* TropSoils Publication, Box 7113, NC State Univ., Raleigh, NC.

Edwards, W. M. 1991. Soil Structure: Processes and Management. In *Soil Management for Sustainability,* pp.7–14. R. Lal and F. J. Pierce, Eds. Soil Water Conserv. Soc. Am.

Follett, R. H. and R. F. Follett. 1983. Soil and Lime Requirement Tests for the 50 States and Puerto Rico. *J. Agron. Educ.* 12:9–17.

Foth, H. D. 1990. *Fundamentals of Soil Science.* 8th ed. John Wiley & Sons, New York.

Foy, C. D. and G. D. Burns. 1964. Toxic Factors in Acid Soils. *Plant Food Rev.* 10:1–2, No. 3. Nat. Plant Food Inst., Washington, D.C.

Hargrove, W. L. and G. W. Thomas. 1981. Effect of Organic Matter on Exchangeable Aluminum and Plant Growth in Acid Soils. In *Chemistry in the Soil Environment.* Am. Soc. Agron. Spec. Pub. 40, Madison, WI.

Haynes, R. J. 1984. Lime and Phosphate in the Soil-Plant System. *Adv. Agron.* 37:249–315.

Kamprath, E. J. 1972. Soil Acidity and Liming. In *Soils of the Humid Tropics*. Nat. Acad. Sci., Washington, D.C.

Kamprath, E. J. 1984. Fertility Management of LAC Soils. *Agron. Abstracts*. Am. Soc. Agron., Madison, WI.

Kamprath, E. J. and C. D. Foy. 1985. Lime-Fertilizer Interactions in Acid Soils. *Fertilizer Technology and Use*. Am. Soc. Agron., Madison, WI.

Krug, E. E. and C. R. Frink. 1983. Acid Rain on Acid Soil: A New Perspective. *Science* 221:520–525.

Lucas, R. E. 1982. Histosols. *Res. Report 435*. Mich. State Agr. Exp. Sta., East Lansing, MI.

Lucas, R. E. and J. F. Davis. 1961. Relationships Between pH Values of Organic Soils and Availabilities of 12 Plant Nutrients. *Soil Sci.* 92:177–182.

Manrique, L. A., C. A. Jones and P. T. Dyke. 1991. Predicting Cation-Exchange Capacity from Soil Physical and Chemical Properties. *Soil Sci. Soc. Am. J.* 55:787–794.

Marion, G. M., D. M. Hendricks, G. R. Dutt and W. H. Fuller. 1976. Aluminum and Silica Solubility in Soils. *Soil Sci.* 121:76–82.

McLean, E. O. 1976. Chemistry of Soil Aluminum. *Commun. in Soil Science and Plant Analysis* 7(7):619–636.

Moberg, J. P, I. E. Esu and W. B. Malgwi. 1991. Characteristics and Constituent Composition of Harmattan Dust Falling in Northern Nigeria. *Geoderma* 48:73–81.

Myers, J. A., E. O. McLean and J. M. Bigham. 1988. Reductions of Exchangeable Magnesium and Liming of Acid Ohio Soils. *Soil Sci. Soc. Am. J.* 52:131–136.

North Central Region Soil Testing Committee-13. 1975. Recommended Chemical Soil Test Procedures. *North Dak. Agr. Exp. Sta. Bul. 499*. Fargo, ND.

Olsson, M. and Per-Arne Melkerud. 1989. Chemical and Mineralogical Changes During Genesis of a Podzol from Till in Southern Sweden. *Geoderma* 45:267–287.

Rogers, H. T., F. Adams and D. L. Thurlow. 1953. Lime Needs of Soybeans on Alabama Soils. *Alabama Agric. Exp. Sta. Bull. 452*.

Russell, J. S., E. J. Kamprath and C. S. Andrew. 1988. Phosphorus Sorption of Subtropical Acid Soils as Influenced by the Nature of the Cation Suite. *Soil Sci. Soc. Am. J.* 52:1407–1410.

Sanchez, P. A. 1976. *Properties and Management of Soils in the Tropics*. John Wiley & Sons, New York.

Sanchez, P. A., J. H. Villachica and D. E. Bandy. 1983. Soil Fertility Dynamics After Clearing a Tropical Rainforest in Peru. *Soil Sci. Soc. J.* 47:1171–1178.

Shoemaker, H. E., E. O. McLean and P. F. Pratt. 1961. Buffer Methods for Determining Lime Requirement of Soils With Appreciable Amounts of Extractable Aluminum. *Soil. Soc. Am. Proc.* 25:274–277.

Soil Survey Staff. 1975. Soil Taxonomy. *USDA Agr. Handbook 436*, Washington. D. C.

Soil Survey Staff. 1992. Keys to Soil Taxonomy. 5th ed. *Soil Mgt. and Support Services Monograph 6*, USDA, Washington, D.C.

Spain, J. M., C. A. Francis, R. H. Howeler and F. Calvo. 1975. Differential Species and Varietal Tolerance to Soil Acidity in Tropical Crops and Pastures. *Soil Management in Tropical America*. Soil Sci. Dept., North Carolina State Univ., Raleigh, NC.

Spurway, C. H. 1941. Soil Reaction Preferences of Plants. *Spec. Bul. 306*. Mich. Agr. Exp. Sta., East Lansing, MI.

Sumner, M. E. 1991. Soil Acidity Control Under the Impact of Industrial Society. In *Interactions at the Soil Colloid-Soil Solution Interface*, pp. 517–541. G. H. Bolt et al, Eds. NATO ASI Series Vol. 190. Kluwer Academic Publishers, London.

Thomas, G. W. and W. L. Hargrove. 1984. The Chemistry of Soil Acidity. In *Soil Acidity and Liming*. 2nd ed. F. Adams, Ed. Agronomy 12:1–58. Am. Soc. Agron., Madison, WI.

Wolcott, A. R., H. D. Foth, J. F. Davis and J.C. Shickluna. 1965. Nitrogen Carriers: Soil Effects. *Soil Sci. Soc. Am. Proc.* 29:405–410.

Wolt, J. D. and D. A. Lietzke. 1982. The Influence of Anthropogenic Sulfur Inputs upon Soil Properties in the Copper Basin Region of Tennessee. *Soil Sci. Soc. Am. J.* 46:651–656.

CHAPTER 5

Soil pH Management

Only 11% of the world's soils have no serious limitations for agricultural use. Drought is a major limitation on about 28% and mineral stress is a major limitation on 23%. Many soils that are affected by drought are moderately or strongly alkaline and many soils that are affected by mineral stress are strongly acid. Thus, there is a need to alter or manage the pH of many of the world's soils to make them more productive. Liming of acid soils to increase pH will be considered first, followed by the acidification of alkaline soils.

5.1 THE pH PREFERENCES OF PLANTS

Plants that have evolved in desert regions have adapted to thrive on neutral and alkaline soils, and many desert plants have great tolerance for soluble salts and soluble Na (halophytes). Plants that are native to the humid tropics, such as cassava, have a high tolerance for soluble Al. Rhododendron grows well on highly acid soils and appears to be an Al accumulator plant. Great genetic variation exists between species and cultivars, and this fact is used to develop new varieties that can better cope with stressful soil conditions related to soil pH. For example, moderately Al-tolerant wheat varieties have been developed for production on Oxisols in Brazil. The pH preferences of over 1,500 plants was published by Spurway.

From Table 5.1, it can be noted that legumes, like alfalfa, sweet clover, and soybeans produced maximum yields at a pH of 6.8. The needs of both the legume plants and their N-fixing organisms must be considered. Oats grew quite well at pH 4.7 and about as well at 5 as at 5.7, 6.8, or 7.5. Corn (maize) produced well over the range of 5.0 to 7.5. By contrast, barley failed to yield at 4.7 and attained maximum yield only when the pH was 6.8 or higher. For all crops, the relative average yields at pH 4.7 were only 32% compared to 98% at pH 6.8.

Several large relative-yield decreases occurred when soil pH decreased from 5.7 and this is near the pH when many soils become less than 100% basic cation saturated and the amount of XAl increases rapidly. Sorghum growth on Ultisols

Table 5.1 Relative Crop Yields versus Soil pH

Crop	Relative yield at indicated pH				
	4.7	5.0	5.7	6.8	7.5
Alfalfa	2	9	42	100	100
Barley	0	23	80	95	100
Corn (maize)	34	73	83	100	85
Oats	77	93	99	98	100
Red clover	12	21	53	98	100
Soybeans	65	79	80	100	93
Sweet clover	0	2	49	98	100
Wheat	68	76	89	100	99

Source: Data from Ohio Agr. Exp. Sta. Spec. Cir. 53, 1938.

in Georgia decreased markedly from pH 5.7 to 5.1 and the decreased growth was associated with a marked decrease in basic cations and increase in XAl (Table 5.2).

Table 5.2 Effect of Acidification on Sorghum Yields and Exchange Properties of a Typic Hapludult Soil

Soil pH	CEC at field pH (cmol/kg)	Extractable (cmol/kg)		Basic cations (cmol/kg)	Sorghum yield (tonne/ha)
		Al	H		
4.7	2.72	1.88	0.16	0.61	0.16
5.1	2.89	1.07	0.09	1.49	2.00
5.7	3.10	1.03	0.14	2.64	3.60
6.2	4.62	0.00	0.08	4.66	4.20

Source: Sumner, 1991.

Data of this nature are the basis for recommending lime to increase pH of some acid soils to a particular value. For example, the pH may be 6.0 or less for the production of continuous corn. However, if corn is included in a rotation with alfalfa, the pH should be increased for profitable production of the alfalfa.

5.2 NEUTRALIZATION REACTIONS

The major liming material is ground limestone containing $CaCO_3$ and $MgCO_3$. The carbonates in lime hydrolyze to produce OH^-.

$$CaCO_3 + H_2O = Ca^{2+} + HCO_3^- + OH^- \qquad (5.1)$$

The H^+ in the soil solution of acid soils comes mainly from Al hydrolysis. The H^+ reacts with the OH^- to form water. The overall reaction representing the neutralization of Al derived soil acidity can be written as:

$$2AlX + 3CaCO_3 + 3H_2O = 3CaX + 2Al(OH)_3 + 3CO_2 \tag{5.2}$$

If a soil with a pH less than 5 is limed to pH of 7, there will be a progressive series of neutralization reactions. The exchangeable Al will be neutralized first followed by the hydroxy-Al forms. The Al is in six-coordination with H_2O or OH^- ligands, $H_2O + OH^- = 6$, and, when equations are written with Al fully hydrated, it is easier to follow the changes that occur with pH changes.

$$Al(6H_2O)^{3+} + OH^- = AlOH(5H_2O)^{2+} + H_2O \tag{5.3}$$

$$AlOH(5H_2O)^{2+} + OH^- = Al(OH)_2(4H_2O)^+ + H_2O \tag{5.4}$$

$$Al(OH)_2(4H_2O)^+ + OH^- = Al(OH)_3(3H_2O)^0 + H_2O \tag{5.5}$$

Each reaction is driven to the right as H^+ accumulated during soil acidification is neutralized by the OH^- supplied by the lime. At pH about 7, the hydroxy-Al is precipitated as $Al(OH)_3$. The various forms of Al in solution for neutralization versus pH are given in Figure 4.4. The decrease in Al saturation and increase in pH due to liming are shown in Figure 5.1.

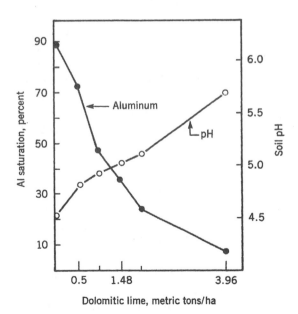

Figure 5.1 The effect of liming on soil pH and the percentage of aluminum saturation of an Ultisol. (Data from the annual report on tropical soils research of North Carolina State University, 1976.)

The neutralization of acidity and the increase in pH result in an increase in the effective cation exchange capacity (ECEC). The ECEC of some low activity clay soils may be increased four or more times by increasing the pH to 7, however, these soils would not normally be limed to pH 7. After the XAl is neutralized and the pH is 5.5, the soil is about 100% XCa..Na saturated and any further increase in pH results in a greater quantity of XCa..Na, because the ECEC increases. There will be a greater supply of available Ca and Mg when lime containing both Ca and Mg is used to neutralize soil acidity. Using lime without Mg and CaO to increase pH of some Ohio soils to pH 7 resulted in 17 to 34% reduction in XMg, which was believed to have been brought about by the occlusion or co-precipitation of Mg with Al.

5.3 THE LIME REQUIREMENT

The lime requirement is *the amount of liming material required to change the soil to a specified state with respect to pH or soluble Al content.* Three different liming strategies or lime requirement methods can be recognized for three different kinds of mineral soil situations: high activity clay soils with high exchangeable basic cations (XCa..Na), low activity clay (LAC) soils with low ECEC and low XCa..Na, and LAC soils with very low ECEC and very low XCa..Na.

5.3.1 Soils with High ECEC and XCa..Na

Most mineral soils, other than Oxisols and Ultisols, have high ECEC and high XCa..Na. Deficiencies of Ca and Mg and aluminum toxicity are seldom a problem. These soils are typically no more than moderately weathered and have a significant content of primary minerals that can weather and supply both macro- and micronutrients. The soils are buffered against changes in availability of micronutrients with moderate changes in soil pH. Overliming, however, may significantly affect availability of micronutrients, commonly the case for Mo. For these soils a considerable change in pH may be associated with little change in yield, as indicated by the data in Table 5.1. Thus, the lime requirement is the amount of lime that will adjust the soil to some selected pH, based primarily on the needs of the crops that are produced.

Theoretically, the best way to determine the amount of lime to add to achieve a particular pH is to add varying amounts of lime and determine the equilibrium pH (Figure 5.2). A rapid method is needed in soil-testing labs where a large number of samples are tested. Commonly, a known amount of buffer solution with a given pH (7, 7.5, or 8.0) is mixed with a standard amount of soil. After a standard period of stirring and standing, the pH of the soil and buffer suspension is measured. The depression of pH from that of the buffer solution is equated with the amount of lime needed to increase soil pH to certain levels as shown in Table 5.3.

The Shoemaker, McLean, and Pratt buffer method (SMP) is recommended for use in the north-central states and is used in about 18 states. The buffer consists

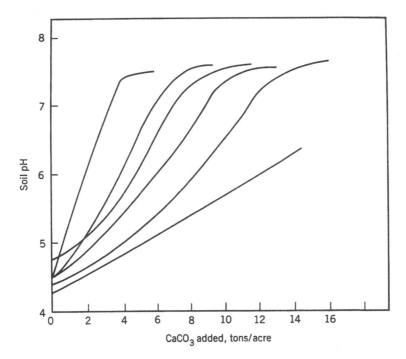

Figure 5.2 Soil pH as a function of added calcium carbonate. The soil on the left is the least buffered and the soil on the right is the most buffered. (Data from Shoemaker, McLean, and Pratt, 1961.)

Table 5.3 Lime Requirement to Achieve Various Soil pH Levels for an 8″ Plow Layer per Acre

| Soil + buffer, pH | Mineral soils, tons of limestone* needed for | | | Organic soils, tons of limestone needed for |
	pH 7.0	pH 6.5	pH 6.0	pH 5.2
6.8	1.4	1.2	1.0	0.7
6.6	3.4	2.9	2.4	1.8
6.4	5.5	4.7	3.8	2.9
6.2	7.5	6.4	5.2	4.0
6.0	9.6	8.1	6.6	5.1
5.8	11.7	9.8	8.0	6.2
5.6	13.7	11.6	9.4	7.3
5.4	15.8	13.4	10.9	8.4
5.2	17.9	15.1	12.3	9.4
5.0	20.0	16.9	13.7	10.5

* Ground agricultural lime with 90% neutralizing power and 40% through 100 mesh, 50% through 60 mesh, 70% through 20 mesh, and 95% through 8 mesh.

Source: Recommended Chemical Soil Test Procedures, North Central Region Soil Testing Committee-13, 1975.

of a combination of p-nitrophenol, triethanolamine, potassium chromate, calcium acetate, and calcium chloride. The buffer agent is adjusted to pH 7.5 with NaOH. This method is well suited for soils with appreciable buffer acidity and considerable XAl. In the southeastern U.S., about four states use the Adams and Evans buffer method. The buffer reagent is a combination of p-nitrophenol, boric acid, potassium chloride, and potassium hydroxide adjusted to pH of 8. The method is well suited to soils with low ECEC and the amount of lime needed is small. Lime requirement of Histosols is similarly determined with buffer solutions; however, liming to no more than pH of 5.2 is recommended because above pH 5.2 the availability of some of the micronutrients is reduced. In addition, the chelation of Al^{3+} and Mn^{2+} greatly reduces the likelihood of toxicity in acid organic soils.

5.3.2 Soils with Low ECEC and XCa..Na

Most of the Ultisols and Oxisols used for cropping are LAC soils with low ECEC and low XCa..Na. There is commonly sufficient ECEC and Al saturation to make Al toxicity the primary limitation in cropping. Aluminum toxicity is likely for moderate and low Al tolerant crops when the Al saturation is 60% of the ECEC, 67% of CEC determined at pH 7, or 86% of the CEC determined at pH 8.2. Manganese toxicity and Ca deficiency also occur frequently. The lime requirement is the amount of lime needed to remove the threat of Al toxicity for Al intolerant cultivars. Experience has shown that this is an amount of lime equivalent to 1.5 times the XAl, and the pH will be increased to about 5.5. This amount of lime will neutralize 85 to 90% of the XAl in soils containing 2 to 7% organic matter. This amount of lime will also inactivate Mn and supply sufficient Ca.

The 1.5 factor accounts for neutralization of some XH released from organic matter and oxidic clays. Some soils need a factor greater than 1.5, depending on organic matter content. Each cmol of Al kg^{-1} of soil requires 1.65 tons of $CaCO_3$ ha^{-1}.

The total and available amount of some micronutrients in LAC soils tend to be very low and yields may be very sensitive to increases in pH that affect availability of micronutrients. Note in Figure 5.3 that a small amount of lime, 0.5 ton/ha, or about the amount needed to neutralize the XAl, increased the yield of all four cultivars. Additional lime sharply decreased yield for two cultivars, but two others were little affected. The LAC soils tend to have a small ECEC and liming may produce a significant increase in the ECEC (see Figure 3.3).

The subsoils of many Ultisols and Oxisols limit root growth through Al toxicity, deficient Ca, or both as illustrated in Figure 5.4. Deep incorporation of lime or the downward movement of lime over a long period of time will help to increase yields by increasing the available water supply. The high cost of applying lime to subsoils and lack of available equipment greatly limit the deep incorporation of lime. Calcium deficiency and Al toxicity commonly limit root growth in subsoils and, therefore, commonly limit crop yields on Ultisols of the southeastern U.S. There are no Oxisols recognized in the continental U.S.

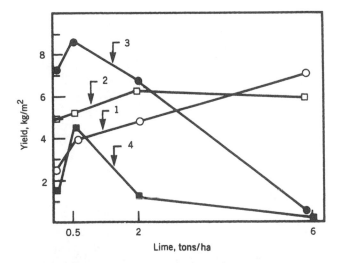

Figure 5.3 The response to liming of four different cassava cultivars grown on Oxisols in Colombia. Type 1, normal response; type 2, little response; type 3, little positive response followed by marked negative effect; type 4, marked initial response followed by drastic negative effect.

5.3.3   Soils with Very Low ECEC and XCa..Na

The most weathered soils, some Oxisols, have very low ECEC and very low XCa..Na and tend to be the most deficient in Ca and/or Mg for cropping. Even though there may be high Al saturation, there tends to be too little XAl to produce enough soluble Al^{3+} to produce toxicity. The lime requirement is the amount needed to correct the deficiency of Ca and/or Mg and is essentially a fertilization problem. The goal is one of adding enough lime to correct the deficiency without adding so much that the availability of some other nutrients, like Zn and B, is reduced enough that the soil becomes deficient in them. Small applications of lime are desirable where lime materials are scarce and must be transported a long distance. The effects of very small lime applications, however, are short lived.

5.4 LIMING PRACTICES

Agricultural lime is a soil amendment consisting principally of $CaCO_3$, and to a lesser extent $MgCO_3$. It is used to neutralize soil acidity and furnish Ca and Mg for the growth of plants. Other liming materials include marl, chalk, wood ashes, blast furnace slag and basic oxygen furnace slag from iron and steel manufacturing, flue dust from cement plants, and refuse from sugar beet and paper mill plants.

Figure 5.4 Shallow rooting of corn caused by acid soil infertility in the subsoil. (Photo courtesy K.D. Ritchey, EMBRAPA (of Brazil) and USAID North Carolina State University Soils Research Project.)

5.4.1 Limestone as a Liming Material

Most of the lime is ground limestone. Limestone containing all $CaCO_3$ is calcitic and limestone containing varying amounts of both $CaCO_3$ and $MgCO_3$ is dolomitic limestone. The two most important properties of ground limestone are neutralizing value and particle size. The neutralizing value determines how much acid can be neutralized by a given amount of limestone and the fineness determines how rapidly the limestone will dissolve and neutralize soil acids.

The neutralizing value of liming materials is easily determined by adding a known amount of lime to a known amount of acid and allowing neutralization to go to completion. The excess acid is titrated with a base, and the neutralizing value is equated to the amount of acid neutralized by the sample. The neutralizing value is expressed as the calcium carbonate equivalent (CCE) with pure $CaCO_3$ rated as 100. If a limestone only contained $CaCO_3$, and no impurities, the CCE would be 100. The CCE of other pure liming materials is given in Table 5.4.

Table 5.4 Neutralizing Value of Liming Materials

Material	Calcium carbonate equivalent
Calcium carbonate	100
Dolomite	109
Calcium hydroxide	136
Calcium oxide	179
Calcium silicate	86

The particle size of limestone is determined by the fineness of grinding. The dissolution of large particles, larger than those able to pass an 8-mesh, is so slow they have little or no affect on soil pH. This fraction has an effective rating of zero. Particles passing an 8-mesh screen but retained on a 60-mesh screen are rated 50% effective. Particles passing a 60-mesh screen are rated 100% effective. It can be observed from Figure 5.5 that particles coarser than 20-mesh are ineffective in increasing soil pH.

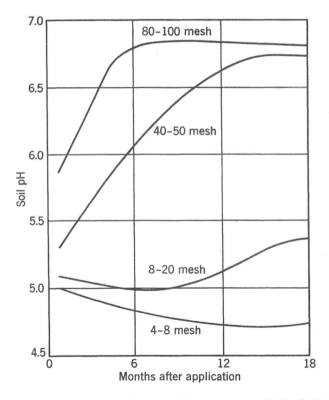

Figure 5.5 Effect of dolomitic lime of different particle sizes on the pH of soil at various times after application. (Adapted from Meyer and Volk, 1952. Used by permission of Williams & Wilkins Co.)

The CCE and the calculated fineness factor are multiplied together to determine the effectiveness of a ground limestone or the effective calcium carbonate

(ECC). If an entire sample passes the 8-mesh screen and 50% is retained on the 60-mesh screen, the fineness factor will be 75. For a sample with a CCE of 90, the ECC will be 67.5.

5.4.2 Lime Application

Major considerations in the application of lime include the availability of the lime and the convenience of spreading. Lime is usually broadcast on the soil surface before tillage operations so that the soil, and limestone are mixed to increase soil and lime contact. Lime can be applied at any time between the harvest of one crop and the planting of the next. Good results have been obtained on LAC soils with planting soon after lime application if the lime is well mixed with the plow layer. Here, the lime reduces soluble Al and increases soluble Ca quite rapidly. Where a significant increase in soil pH is desired before planting on high activity clay soils, lime should be applied several months before planting.

Calcium carbonate is quite insoluble, dissolves slowly, and has very limited downward mobility in soils. Calcitic lime, however, dissolves more rapidly than dolomitic lime. The rate of neutralization is importantly related to fineness, uniformity of distribution in the soil, and the rate of diffusion of Ca from the lime particles to the sites of neutralization. A reasonable diffusion rate for Ca^{2+} is 0.35 cm in 100 days or 0.78 cm in 500 days. If limestone is applied at rate of 6 tons/ha (2.7 tons/acre) in particles 0.60 mm diameter (30-mesh), the average distance between particles of lime will be 1.04 cm. This statistic illustrates the importance of fineness and uniform distribution.

In no-till systems there is limited mixing of soil by tillage operations, and surface applications of N fertilizers will tend to cause rapid acidification of the upper 10 cm of soil. No-till tillage methods provide limited opportunity for incorporation and downward movement of lime to increase the pH of underlying soil layers.

Recent interest in water flow through macropores, causing short circuiting of the fine-pore soil matrix, has directed attention to the role of earthworm burrows in the downward transport of water and solutes. After four lime applications in twenty years of no-till, there was a marked increase in soil pH from below the depth of the original plow layer to a depth of about 70 cm, as compared to conventional tillage (Figure 5.6). This was observed to be the approximate depth of *Lumbricidae terrestris* burrows in the soils that are Alfisols in southeastern Ohio.

The depth of soil sampling and application of lime will need to be adjusted to reflect the particular management system used.

Liming, in a sense, reverses the process that make soils acid. It is only natural to expect that once an acid soil has been limed, leaching and crop production practices will eventually lower the pH again. Studies conducted in Ohio showed the annual loss of Ca by leaching was greater than the Ca removed in harvested crops. When crop rotations include a legume crop, liming of acid soils is required every three to five years in Illinois to maintain a satisfactory soil pH.

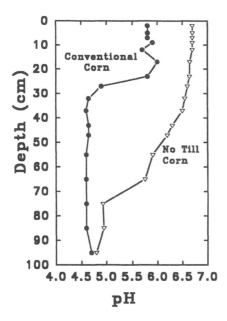

Figure 5.6 Effect of no-tillage and conventional tillage in a corn rotation on the pH with depth of a Westmoreland silt loam. (Adapted from Edwards, 1991. Used by permission of Soil and Water Conservation Society.)

5.4.3 Overliming

Overliming can be a problem when soils have low CEC and a very small amount of buffer or nonexchangeable acidity and the pH can be increased easily. Overliming is most likely on soils with low ECEC, such as sandy soils with permanent charge clays and LAC soils. Plant response to overliming is highly variable and depends on cultivar. Overliming is expensive and injury to crops results primarily from changes in the availability of certain micronutrients. Overliming sometimes produces deficiencies of B, Cu, Mn, and Zn and sometimes produces a Mo toxicity.

5.4.4 Effect of Flooding on Lime Needs

Lime requirements for upland rice are determined similarly as for other crops. Flooded or paddy rice, however, rarely needs lime because the consumption of H^+ during chemical reduction in anaerobic soil increases soil pH. In fact, most paddy soils when flooded have an equilibrium pH between 6.5 and 7.5.

When the O_2 supply of the soil is exhausted, soil microorganisms are forced to use progressively weaker electron acceptors. After O_2, the next strongest electron acceptor available is nitrate, followed by MnO_2, $Fe(OH)_3$, and sulfate. Two examples of the reaction are

$$2NO_3^- + 10e^- + 12H^+ = N_2 + 6H_2O \qquad (5.6)$$

$$Fe(OH)_3 + e^- + 4H^+ = Fe(OH)_2 + OH^- \qquad (5.7)$$

In flooded rice fields, iron is commonly the most abundant electron acceptor in acid soils. Iron reduction increases soil pH owing to the consumption of H^+ and the production of OH^-. The iron supply is much more limited in alkaline soils and the production of carbonic acid from organic matter decomposition has an acidifying effect. Thus, the flooding of most rice fields results in a pH near neutral and liming is a minor aspect of paddy rice production.

5.5 OTHER LIMING MATERIALS

Although ground limestone is the major liming material used in agriculture, there are numerous materials that neutralize soil acidity. These include marl, agricultural slags, and wood ashes.

5.5.1 Marl

Marl is soft and unconsolidated $CaCO_3$, usually mixed with varying amounts of clay and other impurities. Marl occurs as an amorphous and soft deposit found in poorly drained and low-lying areas. It is soft when dry and finely divided. It is dug out and generally spread as chunks rather than fine particles. Some states require that the CCE be 80, while in others the CCE requirement is 70. About 2 m^3 of marl are considered the equivalent to 1 ton of ground limestone.

5.5.2 Blast Furnace and Basic Oxygen Furnace Slag

Blast furnace slag is a byproduct of pig iron production. Iron ore and coke, along with limestone, are loaded into a blast furnace and heated to about 1900°C. This results in a separation of the iron from the silicates and alumina which combine with limestone and form a slag. The molten slag is removed from the furnaces and rapidly cooled, resulting in an amorphous lightweight material in the form of porous granules.

Basic oxygen furnace slag is a later development in the steel-making industry. The steel is made much more rapidly with injection of oxygen into the process, so more calcium oxide is added than is necessary to form the calcium silicates. Consequently, the excess CaO carries over into the slag and produces a faster acting, higher CCE slag than the blast furnace slags.

Slag is mainly Ca and Mg aluminosilicates, $CaSiO_3$ and $MgSiO_3$. When added to acid soils, the silicates neutralize acidity similarity to limestone. The CCE usually ranges from 80 to 100. The use of slag is importantly related to nearness to the slag source.

5.5.3 Wood Ashes

Wood ashes are an ancient soil amendment. Recent environmental and sustainable agriculture concerns have created renewed interest in the application of wood ashes to the land. Burning of wood and related wastes, to generate electricity has created an ash disposal problem. About 90% of the ash in the U.S. is disposed in landfills, although in the Northeast up to 80% of the wood ashes are spread on agricultural land.

Wood ashes generally have a high pH and are used primarily for liming acid soils. This means of disposal can be both environmentally and economically attractive. Recycling of the nutrients contribute to a more sustainable agricultural system.

The high temperature of burning results in the formation of oxides of alkali and alkaline earth elements. During storage, hydration converts the oxides to hydroxides that are subsequently converted to carbonates through absorption of CO_2. Wood ashes are generally high in Ca, Mg, and K carbonates. Their neutralizing value ranges from about 30 to 70% of that of pure calcium carbonate.

In a Maine study, the CCE of wood ashes ranged from 26 to 59% and the predicted CCE was related to the total Ca content.

$$CCE = 0.168 \text{ total Ca} + 13.43$$

There is also a significant amount of P and K in wood ashes. They are about equivalent to fertilizers in the range of 0-1-3 to 0-3-9. Selected properties of the wood ashes are given in Table 5.5. The availability of the K ranged from 39 to 82% and the availability of the P from 43 to 56%.

Table 5.5 Properties of Wood Ashes

Source of ash	CCE	Total nutrient content (g/kg)					
		K	P	Ca	Mg	Na	Al
Ultrapower	56	26.8	5.9	240	11	0.43	2.4
Greenville stream	26	27.2	5.1	74	9	2.18	17.0
Beaverwood chester	53	73.8	14.3	280	22	5.6	6.4
SLASH	50	39.0	12.9	220	20	2.56	14.6
SDWBioash	25	38.8	5.9	88	9	2.72	32.0
Pinetree	59	60.0	12.8	220	20	2.12	5.3

Source: Ohno and Erich, 1990.

5.6 SELECTION OF A LIMING MATERIAL

Liming materials are normally selected by a farmer based on economics. Since transportation is a major part of the cost, distance from the source becomes all-important. Adjustments in time of application may need to be made if the most economical source is coarser sized or is known to be a slower reaction material.

A determination should be made of the need for Mg before selecting a liming material because this is generally the most economical carrier for Mg if liming is necessary. In this case, a premium paid for dolomitic limestone is a savings in the total management of a farm.

5.7 CULTURALLY PRODUCED SOIL ACIDITY

Agricultural practices that tend to increase soil acidity include (1) the addition of H^+ from nitrification of ammonia in fertilizers, farm manures, and legume-fixed N, and (2) the loss of basic cations in harvested crops. Nitrogen fertilizer use and the removal of basic cations in harvested crops has produced large areas of acid soils where normally the soils are alkaline or neutral. Intense acidification occurred in the orchards of Washington as early as 1951 where soils are naturally low in S and ammonium sulfate was the preferred N fertilizer. Lime use is essential to maintain agricultural sustainability.

5.7.1   Acidity and Basicity of Fertilizers

Fertilizers as soluble salts are acid, neutral, or alkaline. Most of the net or long-term effect of fertilizer on soil pH results from transformations that occur in the soil after application. When urea is hydrolyzed to NH_4^+, the pH may increase to 9 or more in the immediate vicinity of application. Later, the NH_4^+ may be absorbed by roots and removed from the soil or nitrified in the soil with the production of protons.

$$2NH_4^+ + 4O_2 = 2NO_3^- + 4H^+ + 2H_2O \qquad (5.8)$$

Whether N is taken up by roots as nitrate or ammonium affects the amount of H^+ or OH^- excreted by roots. Leaching of nitrate resulting from fertilizer use causes a depletion of XCa..Na. The effects of fertilizers on soil pH are complex because the outcome is influenced by crop yields, harvesting methods, soil, and leaching. Pierre did much of the early work and based his conclusions of the effect of fertilizers on soil pH in greenhouse studies. His work forms the primary basis of the method adopted by the Association of Official Agricultural Chemists to determine the acidity or basicity of fertilizers.

The acidity or basicity of a fertilizer is expressed as units of $CaCO_3$ equivalent to the acidity or basicity produced by the fertilizer. It is assumed that (1) Cl, S, and one-third of the N contribute to soil acidification; (2) that the Ca, Mg, K, and Na cations contribute to soil basicity; and (3) that one-half of the N is absorbed by plants as nitric acid and half is absorbed as a salt, such as KNO_3. Thus, KNO_3 and $Ca(NO_3)_2$ are basic and KCl is neutral. The N carriers with an acidic anion, such as ammonium sulfate, are very acid forming. In general it is the N carriers that affect soil pH because of their acidity potential as shown in Table 5.6.

Table 5.6 Equivalent Acidity (A) or Basicity (B) of Selected Fertilizers

Fertilizer	Grade	A or B	Equivalent CaCO$_3$ per unit of N*
(NH$_4$)$_2$SO$_4$	20.5-0-0	A	107
NH$_4$NO$_3$	33.5-0-0	A	36
NH$_3$	82-0-0	A	36
Urea	46-0-0	A	36
N solution	32-0-0	A	36
Ca(NO$_3$)$_2$	16-0-0	B	22
KNO$_3$	13-0-44	B	40
(NH$_4$)$_2$HPO$_4$	11-48-0	A	36
Superphosphate		Neutral	0
KCl	0-0-60	Neutral	0

* Per 20 pounds or 9 kg of N.

Source: Data from Terman, 1982.

A single application of a NPK fertilizer is unlikely to have any detectable long-term effect on soil reaction. The effects depend on rates of application and soil pH and buffering capacity. Long-term addition of ammonium fertilizers for cereal grain production, however, may severely lower soil pH (Figure 5.7).

Figure 5.7 Continued use of ammonium sulfate has acidified soil and produced toxic conditions.

A major limitation of the acidity values is that there are great variations in the fate of fertilizer N due to variations in losses by leaching, denitrification, and volatilization. Some acid-forming fertilizers may produce beneficial effects when used on alkaline or calcareous soils by increasing the solubility of some micronutrients. A quantitative measurement of acidity and basicity enables fertilizer manufacturers to add some limestone to the fertilizer as a filler to counteract acidity. It is generally more economical to add limestone to soils, as compared to the production of neutral fertilizers, if soil acidity becomes a problem due to fertilizer use.

5.7.2 Culturally Produced Acid Soils of the Pacific Northwest

Many naturally acid soils, Andisols and Ultisols, occur in the Coastal and Cascade Ranges of the northwestern U.S. By contrast, the soils east of the mountains at lower elevations on the Plains tend to be neutral or alkaline and naturally low in available S. Sulfur was supplied by the use of sulfur-containing fertilizers. Intense soil acidification, culturally produced, was first observed in Washington orchards about 1950 where coarse-textured soils had been heavily fertilized with ammonium sulfate. Corrective practices include liming, a shift to nitrate N sources, especially $Ca(NO_3)_2$, and foliar applications of urea.

More recently, rapidly increasing acidity of the plow layers of dryland wheat soils has occurred in eastern Washington and northern Idaho. The wheat-producing soils are mainly Mollisols and Xerolls. Use of ammonium fertilizers and increased removal of basic cations are believed to be the major causes of the culturally produced acidity.

Soil acidification under irrigation is greatly affected by the alkaline carbonate content of the water. The variable nature of the irrigation water results in great variation in the effects of cultural practices.

The agriculturally acidified soils of the region appear to have a lower fraction of the exchangeable acidity as Al^{3+}, as compared to geologically acid soils. At pH 4.5, less than 50% of the exchange acidity was XAl, compared to over 90% in geologically acid soils. It appears that there is a slow release of Al from mineral decomposition in the agricultural soils and reduced activity of Al due to reaction with P from high levels of P fertilization.

Cultural practices have also increased the acidity of naturally acid soils in the Pacific Northwest. Grass seed fields in the Willamette Valley have pH values as low as 4.5 where N fertilizers have been used for thirty to forty years.

5.7.3 Culturally Produced Acid Soils of the Great Plains

For many decades, declining soil organic matter (SOM) content contributed a significant amount of N for wheat and forage production. This was followed by a marked increase in the use of N fertilizers on the Great Plains by the 1950s. The pH of the plow layer of Mollisols (Argiustolls) declined from a pH of 6.5 to 4. Wheat failed to grow in some cases.

Yield responses to lime application for wheat occur on very acid soils in south-central Kansas and north-central Oklahoma and Al-tolerant cultivars are used. Generally, lime is recommended to maintain soil pH at 5.5 or higher. Significant yield increases have been obtained with banded P fertilizer. Applying

P fertilizer with the wheat seeds at planting reduces Al toxicity near the seeds. As seedlings age, their roots become more tolerant of Al and the roots invade soil below the plow layer where the pH is higher and soluble Al is very low.

Basic cation removal is an important factor in culturally produced soil acidity. Basic cation removal is affected by crop harvesting methods. For example, if harvesting a crop results in removing most of the N, P, S, and Cl taken up by the crop while leaving residues in the field that contained most of the Ca, Mg, K, and Na, there would be a minimum amount of acidifying elements left in the field for the production of acidity and a maximum amount of basic cations left in the field to contribute to alkalinity. For cereal crops in general, the grain is relatively high in N, P, and S and low in basic cations while, the reverse is true of the straw.

The effect of harvesting method on soil pH is expressed with the excess base/N ratio (EB/N ratio).

$$(Ca + Mg + K + Na) - (Cl + S + P + NO_3 - N)/total \ N$$

Wheat grain, relative to straw, is high in N, P, S, and Cl and low in Ca, Mg, K, and Na producing a low EB/N ratio, less than 1. Conversely, straw has a high ratio, greater than 1. The data in Table 5.7 give predicted EB/N ratios for several harvesting winter wheat systems using fertilizer containing 200 pounds of ammonium N per acre. Harvesting the grain and leaving the straw has a predicted EB/N ratio of 0.03, meaning minimum residual acid forming elements left in the soil and a pure lime need of only 11 pounds per acre. The EB/N ratio for harvesting only the straw is 1.20, resulting in addition of a large amount of acidifying elements and a pure lime requirement of 432 pounds per acres. Harvesting both grain and straw results in an estimated need for 130 pounds of pure lime. Grazing tends to decrease lime needs, compared to harvesting of grain and straw, because the forage is relatively high in acid forming elements.

Table 5.7 Effect on Ammoniacal N and Harvest on Soil Acidity in a Winter Wheat Production System

Crop removal system*	Yield of dry matter (pounds/acre)	Pounds per acre removed		EB/N (meq/100 g)	Theoretical potential acidity
		N	Excess bases		
Grazing	2,000	80	65	0.23	23
Grain	3,600	72	10	0.03	3
Straw	4,000	28	118	1.20	120
Grazing + grain	5,600	152	75	0.13	13
Grain + straw	7,600	100	128	0.36	36
Grazing + grain + straw	9,600	180	193	0.30	30

* All systems received 200 pounds N/acre.

Source: Westerman, 1992.

5.8 SOIL ACIDIFICATION

Plants that prefer quite acid soil include rhododendron, blueberry, azalea, cassava, and tea (Figure 5.8). By contrast many plants, including roses, pin oak, and maple commonly develop Fe or Mn chlorosis when growing on alkaline soil. Sorghum and corn commonly develop chlorosis due to iron and/or zinc deficiency. Scab disease of Irish potatoes, caused by an actinomycete, has been controlled by maintenance of an acid soil in which the organism is inactive.

Figure 5.8 Tea is a calcifuge and dislikes soils containing free calcium. Tea grows well on soils that have a pH of 4.5 and a high percentage of exchangeable aluminum saturation.

Alkaline soil conditions are caused naturally and by soil disturbance during construction. Soil alkalinity is common in arid and semi-arid areas due to limited precipitation, plant growth, and leaching. Many soils of the humid regions have acid horizons underlain by alkaline layers and the mixing or disturbance of soils during construction commonly produces land with alkaline surface layers. In addition, alkaline runoff from alkaline surface soils or $CaCO_3$-containing structures (sidewalks, roads, etc.) alkalize adjacent acid soils. At the Morton Arboretum near Chicago, alkaline runoff was credited for the alkalizing soil and eventually causing Mn stress and decline in a white oak stand located downslope from a large parking lot, The soil pH changes are shown in Figure 5.9.

5.8.1 Acidifying Soil Using Sulfur

Generally, large areas of land are not acidified in the way that liming is routinely applied to agricultural large fields. The most frequent acidification occurs in nurseries and horticultural situations. Sulfur and sulfur compounds are

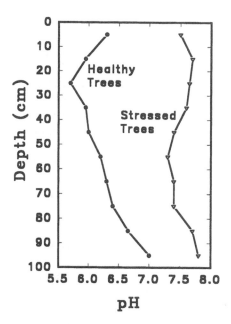

Figure 5.9 Soil pH profiles around stressed and unstressed mature white oaks. (Adapted and redrawn from Messenger, 1986.)

the most popular acidifying compounds. Elemental S is the most economical and is oxidized to sulfuric acid by bacteria, mainly *Thiobacillus*.

$$S + 3O_2 + 2H_2O = 2H_2SO_4 \qquad (5.9)$$

Microbial catalysis promotes the reaction, which is irreversible and occurs slowly. Aluminum (and Fe) sulfate is commonly used for growing garden and landscape plants. The reaction is

$$Al_2(SO_4)_3 + 6H_2O = 2Al(OH)_3 + 3H_2SO_4 \qquad (5.10)$$

The $Al(OH)_3$ precipitates, leaving the acid to acidify the soil. Some of the $Al_2(SO)_4$ may react with the soil to produce additional acidification:

$$Al_2(SO_4)_3 + 2CaHX = Al_2X_3 + 2CaSO_4 + H_2SO_4 \qquad (5.11)$$

The $CaSO_4$ in gypsum has been found to be effective in increasing sorghum yields on calcareous soils in Kansas. The major benefit appeared to be increased availability of Zn. Calcareous soils are not usually made acid because costs are high. When soils are leached of salts, the requirements for S are reasonable, allowing for economical soil acidification. Sometimes S is applied in a band as a fertilizer to increase acidity in a small fraction of the soil mass where early root growth occurs. The precise quantity of S necessary to lower pH is difficult to predict because it depends upon soil texture, organic matter content, and the quantity of free calcium carbonate in a given soil. Suggested rates of S have been given by Jones (1982) showing that when broadcast, a clay soil will require about twice as much finely ground sulfur to lower the pH from 7.5 to 6.5 (800–1000 compared to 400–600 kg/ha). When the soil pH was 8.0, the clay soil required about 1.5 times more S to lower the pH to 6.5 than the sandy soil (1500–2000 compared to 1000–1500 kg/ha). Banding, compared to broadcasting, reduced the amount of S required to lower the pH by more than 50%.

Acid organic materials may be beneficial, but they not practical where a significant decrease in soil pH is desired. Most N fertilizers, ammoniacal, increase soil acidity.

5.8.2   Other Strategies for Growing Plants on Alkaline Soils

Crops can be selected that grow well on alkaline soil. Lime is generally necessary to economically produce alfalfa in humid regions, whereas alfalfa produces abundantly on alkaline soils and is one of the most popular crops in the irrigated regions of the western U.S. Another strategy is to add naturally deficient nutrients to the soil in a fertilizer. An example is Zn fertilization of white beans grown on the calcareous soils of the lake plain of southeastern Michigan. A third strategy is to add deficient nutrients directly to plants. This has particular value where the deficient nutrients react quickly with the soil and are fixed. Nutrients are supplied by foliar sprays and, in the case of trees, as trunk implants.

5.9 SUMMARY

The management of soil pH is focused on the selection of crops that grow well at the existing pH or altering soil pH by liming acid soils and acidifying alkaline soils.

1. The lime requirement of acid soils is an amount of lime required to:
    a. adjust soil pH to a desired value on high activity clay soils and Histosols
    b. inactivate soluble Al on low activity clay soils with high soluble Al
    c. add Ca and/or Mg on low activity clay soils low in soluble Al
2. Most lime is ground limestone. Other liming materials include marl, blast furnace and basic oxygen furnace slag, wood ashes, and a wide variety of other refuse materials.
3. Lime may be applied any time to low activity clay soils where the lime is used to inactivate soluble Al and/or add Ca and Mg. Lime is best applied several months before planting on high activity clay soils and Histosols where lime is used to adjust soil pH.
4. Nitrogen fertilizer use and basic cation removal by cropping have greatly increased the acidity of both naturally acid and alkaline soils, resulting in increased lime needs.

5. Flooding soils, for rice production for example, increases the pH of acid soils and decreases the pH of alkaline soils, resulting in little or no need for lime.

6. Sulfur is applied to alkaline soils to increase their acidity. Other strategies for managing alkaline soils include crop selection and the application of deficient micronutrients to soils or directly to plants.

REFERENCES

Adams, F. 1984. Crop Response to Lime in Southern United States. In *Soil Acidity and Liming*. Agronomy 12, 2nd ed. F. Adams, Ed. Am. Soc. Agronomy, Madison, WI.

Barber, S. A. 1984. Liming Materials and Practices. In *Soil Acidity and Liming*. Agronomy 12, 2nd ed. F. Adams, Ed. Am. Soc. Agronomy, Madison, WI.

Barber, S. A. 1984. *Soil Nutrient Bioavailability*. John Wiley & Sons, New York.

Burmester, C. H., J. F. Adams and J. W. Odom. 1988. Response of Soybean to Lime and Molybdenum on Ultisols in Northern Alabama. *Soil Sci. Soc. Am. J.* 52:1391–1394.

Edwards, W. M. 1991. Soil Structure: Processes and Management. In *Soil Management for Sustainability*, pp. 7–14. R. Lal and F. J. Pierce, Eds. Soil and Water Conservation Society Am.

Follett, R. H. and R. F. Follett. 1983. Soil and Lime Requirement Tests for the 50 States and Puerto Rico. *J. Agron. Educ.* 12:9–17.

Foy, C. D. and G. D. Burns. 1964. Toxic Factors in Acid Soils. *Plant Food Rev.* 10:1–2. No. 3. Nat. Plant Food Inst., Washington, D.C.

Haynes, R. J. 1984. Lime and Phosphate in the Soil-Plant System. *Adv. Agron.* 37:249–315.

Jackson, T. L. and H. M. Reisenhauer. 1984. Crop Response to Lime in Western United States. In *Soil Acidity and Liming*. Agronomy 12, 2nd ed. F. Adams, Ed. Am. Soc. Agronomy, Madison, WI.

Jones, U. S. 1982. *Fertilizers and Soil Fertility*. 2nd ed. Prentice-Hall, Reston, VA.

Kamprath, E. J. 1984. Crop Response to Lime in the Tropics. In *Soil Acidity and Liming*. Agronomy 12, 2nd ed. Am. Soc. Agronomy, Madison, WI.

Kamprath, E. J. and C. D. Foy. 1985. Lime-Fertilizer Interactions in Acid Soils. *Fertilizer Technology and Use*. Am. Soc. Agronomy, Madison, WI.

Lathwell, D. J. and W. S. Reed. 1984. Crop Response to Lime in Northeastern United States. In *Soil Acidity and Liming*. Agronomy 12, 2nd ed. F. Adams, Ed. Am. Soc. Agronomy, Madison, WI.

Lucas, R. E. 1982. Histosols. *Res. Report 435*. Mich. State Agr. Exp. Sta., East Lansing, MI.

McLean, E. O. and J. R. Brown. 1984. Crop Response to Lime in the Midwestern United States. In *Soil Acidity and Liming*. Agronomy 12, 2nd ed. F. Adams, Ed. Am. Soc. Agronomy, Madison, WI.

Manrique, L. A., C. A. Jones and P. T. Dyke. 1991. Predicting Cation-Exchange Capacity from Soil Physical and Chemical Properties. *Soil Sci. Soc. Am. J.* 55:787–794.

Messenger, S. 1986. Alkaline runoff, soil pH and white oak manganese deficiency. *Tree Physiology* 2:317–325.

Meyer, T. A. and G. W. Volk. 1952. Effect of Particle Size of Limestone on Soil Reaction. *Soil Sci.* 73:37–52.

Myers, J. A., E. O. McLean and J. M. Bigham. 1988. Reductions of Exchangeable Magnesium and Liming of Acid Ohio Soils. *Soil Sci. Soc. Am. J.* 52:131–136.

North Central Region Soil Testing Committee-13. 1975. Recommended Chemical Soil Test Procedures. *North Dak. Agr. Exp. Sta. Bul. 499*. Fargo, ND.

Ohno, T. 1992. Neutralization of Soil Acidity and Release of Phosphorus and Potassium by Wood Ash. *J. Environ. Qual.* 21:433–438.

Ohno, T. and M. S. Erich. 1990. Effect of Wood Ash Application on Soil pH and Soil Test Nutrient Level. *Agriculture, Ecosystems and Environment* 32:223–239.

Russell, J. S., E. J. Kamprath and C. S. Andrew. 1988. Phosphorus Sorption of Subtropical Acid Soils as Influenced by the Nature of the Cation Suite. *Soil Sci. Soc. Am. J.* 52:1407–1410.

Sanchez, P. A. 1976. *Properties and Management of Soils in the Tropics.* John Wiley & Sons, New York.

Sanchez, P. A., J. H. Villachica and D. E. Bandy. 1983. Soil Fertility Dynamics After Clearing a Tropical Rainforest in Peru. *Soil Sci. Soc. J.* 47:1171–1178.

Shoemaker, H. E., E. O. McLean and P. F. Pratt. 1961. Buffer Methods for Determining Lime Requirement of Soils With Appreciable Amounts of Extractable Aluminum. *Soil. Soc. Am. Proc.* 25:274–277.

Soil Survey Staff. 1992. Keys to Soil Taxonomy. 5th ed. *Soil Mgt. and Support Services Monograph 6,* USDA, Washington, D.C.

Spain, J. M., C. A. Francis, R. H. Howeler and F. Calvo. 1975. Differential Species and Varietal Tolerance to Soil Acidity in Tropical Crops and Pastures. *Soil Management in Tropical America.* Soil Sci. Dept., North Carolina State Univ., Raleigh, NC.

Spurway, C. H. 1941. Soil Reaction Preferences of Plants. *Spec. Bul. 306.* Mich. Agr. Exp. Sta., East Lansing, MI.

Sumner, M. E. 1991. Soil Acidity Control Under the Impact of Industrial Society. In *Interactions at the Soil Colloid-Soil Solution Interface,* pp. 517–541. G. H. Bolt et al., Eds. NATO ASI Series, Vol. 190. Kluwer Academic Publishers, London.

Terman, G. L. 1982. Fertilizer Sources and Composition. In *Handbook of Soils and Climate in Agriculture,* CRC Press, Boca Raton, FL.

Thomas, G. W. and W. L. Hargrove. 1984. The Chemistry of Soil Acidity. In *Soil Acidity and Liming.* 2nd ed. Agronomy 12:1–58. F. Adams, Ed. Am. Soc. Agron., Madison, WI.

Westerman, R. L. 1992. Factors Affecting Soil Acidity. In *Efficient Use of Fertilizers,* pp. 153–161. Agronomy 92-1, Oklahoma State University.

Wolcott, A. R., H. D. Foth, J. F. Davis and J.C. Shickluna. 1965. Nitrogen Carriers: Soil Effects. *Soil Sci. Soc. Am. Proc.* 29:405–410.

CHAPTER 6

Nitrogen

Nitrogen is a part of all living cells. In plants, N is a constituent of chlorophyll, all proteins including the enzymes, and many other compounds. Of the nutrients removed from the soil by plants, only H atoms are present in a greater number than N atoms. A lack of N causes leaves to become yellow and stunts growth, but conversely with an adequate supply of N, vegetative growth is rapid and foliage dark green in color as shown in Figure 6.1. The large need of plants for N and the limited ability of soils to supply available N cause N to be the most limiting nutrient for crop production on a global basis.

6.1 THE NITROGEN CYCLE

The great bulk of the earth's N is in the rocks and sediments of the lithosphere; however, the most important reservoir of N for plants and animals is the atmosphere. Pure, dry air is about 78% N by volume, resulting in 77,350 metric tons/ha (34,500 tons/acre) of N. Considering the average biological need for N, the atmosphere contains about a million years' supply.

6.1.1   Inventory of the Earth's Nitrogen

About 98% of the world's N exists in the lithosphere (Table 6.1). Nitrogen is a component of coal and many other rocks and minerals. Ammonium is fixed in the clays of sediments and in mica minerals in the same voids where K^+ is fixed. The second largest N reservoir is the atmosphere, which contains about 2% of the total. Most of the N in the atmosphere is believed to have originated from the lithosphere, and even today volcanic gases contribute N to the atmosphere. By contrast, the amount of N in soils is very small, being only 1/5,000th of that of the atmosphere. Many productive mineral soils contain about 4,000 kg/ha (or about 4,000 pounds/acre) in the furrow slice. About 90% of the soil N is unavailable in organic matter, and most of the remainder exists as fixed ammonium in clays. At any one instant, about 1% or less of the total N in soils is

Figure 6.1 Circles of dark-green grass surrounding trees caused by application of nitrogen fertilizer.

Table 6.1 Inventory of the Earth's Nitrogen

Sphere	Million metric tons
Lithosphere	1.636×10^{11}
Igneous rocks of crust	1.0×10^9
Igneous rocks of mantle	1.62×10^{11}
Core of the earth	1.3×10^8
Sediments (fossil N)	$3.5–5.5 \times 10^8$
Coal	1.0×10^5
Sea bottom organic compounds	5.4×10^5
Terrestrial soils	
Organic matter	2.2×10^5
Clay-fixed NH_4^+	2.0×10^4
Atmosphere	3.86×10^9
Hydrosphere	2.3×10^7
Biosphere	2.8×10^5

Source: Estimates from various sources as given by Stevenson, 1982.

available to plants and microorganisms as nitrate or exchangeable ammonium. This available fraction is rapidly consumed by plants and microorganisms and susceptible to leaching. Therefore, if it is not replaced by fertilization or mineralization, available N becomes even lower. The amount of N in the biosphere, like that in the soil, is a very small part of the total, but this N is mobile and life depends on it.

6.1.2 Factors Affecting Soil Nitrogen Content

Factors associated with accumulation of N in soils include those that favor plant growth, the major source of soil organic matter, and those factors that inhibit organic matter decomposition. Thus, Histosols have the highest N content and dry, hot, and sandy Aridisols the lowest. Grassland soils contain more organic matter and N than nearby forest soils developed under similar conditions. Other factors favoring organic N accumulation in soils include high clay content and low temperature.

Generally, the organic matter and N content decrease in soils with increasing soil depth. In some soils there is a secondary increase of N in subsoils because of the protection of organic matter in Bt horizons with high clay content. Organic matter accumulates in Bhs horizons, as in Spodosols. Organic matter in subsoils is generally older and more decomposed with a lower C/N ratio than organic matter in the A horizon.

6.1.3 The Soil Nitrogen Cycle

The N cycle in soils is a part of the earth's overall N cycle. The annual rate of removal of N from the atmosphere by fixation and its addition to soils is approximately balanced by an equal amount of N returned to the atmosphere by denitrification. Thus, over time, a quasi-equilibrium exists between the amount of N in the atmosphere and the amount of N in soils on a global basis.

The N from the atmosphere circulates through the soil and eventually returns to the atmosphere through a series of processes: fixation, mineralization, nitrification, immobilization, and denitrification. These processes, as shown in Figure 6.2, can be viewed as a series of irreversible microbial reactions that shuttle N back and forth at the discretion of the microorganisms. Ammonium and nitrate are the forms of N available to plants and microorganisms.

6.1.4 Importance of Biological Nitrogen Fixation

Nitrogen fixation is both biological and nonbiological. Some nonbiological N fixation is caused by lightning discharges. Other N in the atmosphere originates from burning of fossil fuels and forests and from the emission of magmatic gases. This N is added to soils as nitrate and ammonium in precipitation. Most of the N added naturally to soils is from biological fixation that is symbiotic or non-symbiotic in nature. In all ecological niches, there are N-fixing organisms including bacteria, algae, and actinomycetes. It has been estimated that of the 12.8 kg N/ha (11.4 pounds N/acre) of N naturally added to soils annually, 72% comes from biological N fixation and 28% from the other sources. The other sources are mainly from lightning discharges, burning of fossil fuels and forests, and from the emission of magmatic gases. This N is added to soils as nitrate and ammonium in precipitation.

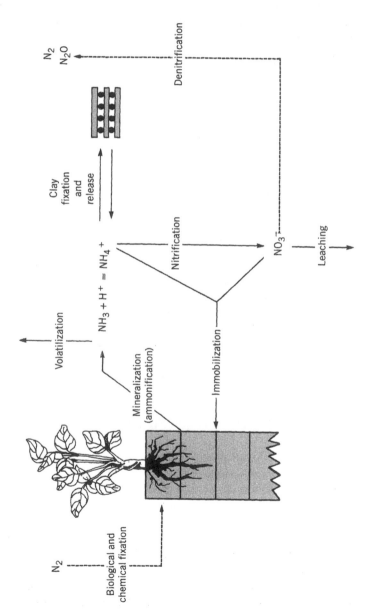

Figure 6.2 The major segments of the soil nitrogen cycle consist of fixation, mineralization, nitrification, immobilization, and denitrification. Some nitrogen is also lost in the cycle by volatization and leaching and some ammonium is temporarily unavailable because it is fixed in clay.

6.2 BIOLOGICAL NITROGEN FIXATION

Considering the large amount of N added to soils by biological fixation and the importance of N in plant growth, biological N fixation can be considered one of the most important processes in nature. In a way, it is similar to photosynthesis. There is a ubiquitous need for N in nature and many different kinds of microorganisms that fix N. The great diversity of fixation sites and organisms is illustrated by a flooded rice field (paddy), as shown in Figure 6.3.

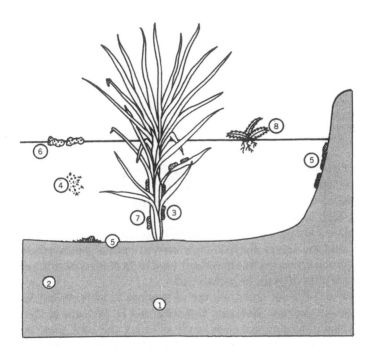

Figure 6.3 Diagram of nitrogen-fixing organisms in a rice field ecosystem. Bacteria: (1) rhizosphere, (2) soil, and (3) epiphytic on rice. Cyanobacteria (blue-green algae): (4) free floating, (5) soil-water interface, (6) water-air interface, (7) epiphytic on rice. Azolla: (8) floating on water surface. (Drawn after Kulasooriya et al., 1980.)

Biological N fixation systems have been grouped into four categories. These systems, representative organisms involved, and amounts of N fixed annually are given in Table 6.2.

6.2.1 Symbiotic Nitrogen Fixation

In symbiotic systems, the host plant supplies the N-fixing organisms with fixed C, a photosynthate, and the host plants benefit from the N fixed. The N-fixing microorganisms are bacteria, actinomycetes, and blue-green algae (cyano-

Table 6.2 Relative Annual Rates of Nitrogen Fixation

System and organisms	N₂ fixed, kg/ha
Symbiotic systems	
Legumes, bacteria	
Alfalfa	128–600
Lupins	150–169
Clover	104–160
Soybeans	57–94
Cowpeas	84
Nonlegume-nodulated, actinomycetes	
Alnus (alder)	40–300
Ceanothus	60
Plant-algal associations	
Lichens	39–84
Gunnera	12–21
Azollas	313
Nonsymbiotic systems, free-living organisms	
Blue-green algae	25
Azotobacter	0.3
Clostridium pasteurianum	0.1–0.5

Source: Data from various sources as given by Evans and Barber, 1977.

bacteria). All of these organisms, which are called diazotrophs, have a very simple cell structure without a nucleus and synthesize the enzyme nitrogenase. Dinitrogen, N_2, is reduced to NH_3 (from valence of 0 to –3). In a chemical plant N is fixed by using high temperature and pressure. The diazotrophs accomplish the same feat at ambient temperature and atmospheric pressure. Biological N fixation is not free because the energy for fixation is derived from a photosynthate. For soybeans, it has been estimated that the energy used to fix N is equal to the energy in about 20 bushels/ha of grain (8 bushels/acre).

Since Greek and Roman times, the use of legumes to increase soil fertility has been a common practice. Hermann Hellriegel and H. Wilfarth of Germany in 1888 found that nonlegumious crops, such as barley and oats, grew in sand culture in direct response to the amount of N supplied. For legumes, however, there was no relation between their growth and the addition of fertilizer N. In this way the link between nodules on legume roots and N-fixation was discovered. Legumes are dicots that develop a symbiotic N-fixation relation with bacteria of the genus *Rhizobium*. It is estimated that over half of the biologically fixed N added to the earth is due to legume symbioses in agricultural production, and this is equal to two times the amount of N added to soils in fertilizers. The importance of legumes in agriculture is shown by their high rates of N fixation (see Table 6.2).

Rhizobia tend to be host specific, although some infect several different hosts, and more than one species may infect a single plant. The bacteria can live in the soil for long periods of time in a nonsymbiotic state. However, to ensure that the proper species is present, farmers usually plant seed inoculated with bacteria that will form an efficient N fixing system with the legume.

Bacteria living near the root of a host plant apparently have a recognition mechanism that triggers a host-bacteria recognition event. The presence of the bacteria adjacent to a root hair causes branching and curling, followed by the invasion of the bacteria. An infection thread is formed that penetrates the root as the bacteria continue to divide and multiply. The plant's response is to form a tumorous nodule containing cells that become packed with *bacteroids*, which are bacteria that have undergone morphological and metabolic change. The bacteroids are supplied with photosynthate, which is used for respiration and N fixing activities. The N fixed as ammonia is excreted from the bacteroids to the legume cells and is then transported as N-C compounds in the vascular system. The relationship is one of true symbiosis, since there is a direct connection between the host and symbiont for the transfer of photosynthate to the symbiont and the transfer of fixed N to the host (Figure 6.4).

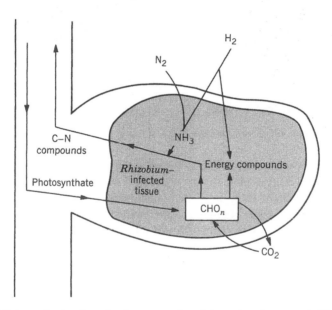

Figure 6.4 Schematic drawing of a legume root nodule and root segment. (From Phillips and DeJong, 1984. Used by permission of the American Society of Agronomy and the Soil Science Society of America.)

Rhizobia form associations with many trees. However, the most important symbiotic relationships with dicotyledonous woody shrubs and trees are those with certain actinomycetes of the *Franki* species. The infection of roots and nodule formation have similarities to the legume-rhizobia systems. *Alnus* is a pioneer genus of large trees that grow in the Rocky Mountains and on the Pacific slope. They are found on freshly exposed parent materials and are an important N contributor in the early stages of soil formation. After clear-cutting of Douglas fir forests in the Pacific Northwest of the U.S., *Ceanothus-Frankia* symbiosis contributes a significant amount of N for the regeneration of the Douglas fir forests (see Table 6.2). Nodules on the roots of *Ceanothus*, a showy genus of shrubs and woody vines, are shown in Figure 6.5.

Figure 6.5 Nodules on the roots of *Ceanothus*, in which actinomycetes fix nitrogen.

One of the best-known symbiotic algal N-fixing associations consists of blue-green algae and fungi in lichens. The algae fix N that benefits the fungi, and the algae appear to benefit from biotin and thiamine produced by the fungi. In wet and tropical environments, blue-green algae invade the stems of herbaceous dicots of the *Gunnera* genus. Glands are formed on the stems, similar to nodules on roots, and the algae fix N in the glands.

The most important symbiotic-algal association for agriculture is the *Azolla-Anabaena* association, which is used for rice production in flooded fields. The azolla is a floating fern with roots that penetrate below the water surface (see Figure 6.3). The anabaenas, which are blue-green algae, live in the fronds of the fern. There is no vascular connection between the two organisms. Up to 60% of the fern's N comes from ammonia produced by the anabaenas. The benefits derived from the association by the algae are unknown. The rapid growth of biomass that is possible under favorable conditions allows for the production of large amounts of green manure rich in N (see Table 6.2). In rice paddies, the N is released to the rice when the organic matter is mineralized. The floating biomass can also be skimmed off the surface of ponds and used as a high-protein animal feed. Thus, great diversity exists in the symbiotic N-fixing systems that account for the addition of large amounts of N to soils in natural ecosystems, as well as in agricultural fields.

6.2.2 Nonsymbiotic Nitrogen Fixation

Several kinds of heterotrophic bacteria are N fixers and are of minor importance in agricultural soils. These heterotrophs must compete with other soil organisms for the limited supply of organic substrates that are available for their C and energy. *Azotobacter* and *Beijernickia* are aerobes and occur in temperate and tropical soils, respectively. *Clostridium* is a heterotrophic bacterium that thrives only under anaerobic conditions. *Azospirillum* is a bacterium that has been found to live in the rhizosphere of the roots of tropical grasses and fix some N.

Blue-green algae live near the soil surface and fix N nonsymbiotically. They are photosynthetic and need not compete for the limited supply of organic substrates. But, because of shading by the soil surface and drying at the soil surface, they are quite inactive in most soils. In rice paddies or other flooded areas, blue-green algae in the water fix large amounts of N and contribute significantly to the N needs of plants.

6.3 MINERALIZATION

All of the mineral N added to soils by fixation and fertilizers are subject to uptake by roots and microorganisms and to conversion into organic soil N. The organic N is reconverted to mineral form by a wide variety of heterotrophic organisms — mainly bacteria and fungi — in a process called *mineralization*. These soil organisms secrete extracellar proteolytic enzymes that decompose proteins. In many well-drained mineral soils, about 2% of the organic N is mineralized annually. For soils with 0.05 to 0.10% N, this amounts to 25 to 50 kg/ha (22 to 45 pounds/acre) for a 20-cm thick plow layer.

Since ammonia is the first mineral form produced, the mineralization of N is also called *ammonification*. Ammonia is a gas and can be lost from the soil by volatilization when mineralization occurs on the soil surface. Much of this NH_3 may be absorbed by leaves if there is a plant canopy. Conversely, near cattle feedlots and certain industrial sites, the atmosphere may be enriched with NH_3 and a significant amount of NH_3 may be adsorbed by nearby soils and lakes. The ammonia molecule is strongly polar and readily combines with a proton to form ammonium, NH_4^+ (see Figure 6.2). Ammonia is stabilized in acid soils but in alkaline soils fewer protons are available and volatilization increases as the soil pH increases. If moisture conditions are favorable (near field capacity), little NH_3 is lost by volatilization even in alkaline soils.

6.3.1 Factors Affecting Mineralization

The factors that affect mineralization are those that affect microbial activity. In general, these are the same factors that affect plant growth including temperature, water content, oxygen availability, pH, supply of nutrients, and salinity.

Temperatures favorable for root growth are generally favorable for N mineralization. Early-growing crops that are planted in wet and cold soils may be stimulated by early N fertilizer applications because of low mineralization. Within the range of 10 to 40°C, a 10-degree increase in temperature increases the mineralization rate two to three times. The optimum temperature for N mineralization in soils appears to be in the range of 30 to 40°C. But in compost piles the thermophiles take over and may be active at temperatures exceeding 65°C.

Mineralization of N is limited in dry soils with very low water potential. As soil becomes wetter, mineralization increases to a maximum, followed by a reduction in rate of mineralization as the soil approaches water-saturation. Mineralization in water-saturated soil is limited by low oxygen availability. Maximum accumulation of mineral N in two weeks in three soils representing a wide range in texture is shown in Figure 6.6. Several observations from this and other studies show maximum mineral N accumulation occurs when soils are slightly wetter than field capacity, water potential from −30 to −10 kp, and with 80 to 90% of the air space filled with water. Mineralization is more depressed near water saturation in clay soils, as compared to sandy soils. The overall plant availability of N occurs in nonwater-saturated soils when mineralization is maximum because mass flow is also high. As soils become drier than field capacity, N availability decreases due to decreasing N mineralization and decreasing movement of N to roots by mass flow.

The appreciable amount of N mineralization in soils drier than the wilt point, −1,500 kp, suggests that an appreciable amount of available N may accumulate in soils during drought and contribute significantly to plant growth when favorable conditions are resumed.

Wetting of dry soils appears to stimulate mineralization and brings a period of rapid mineralization or a *flush* in N availability. The longer the dry period, the stronger is this effect which has not been satisfactorily explained. It has been suggested that rapid mineralization in savannah soils after a long dry period results in much NO_3^-, which increases the cation concentration in the soil solution to maintain charge balance. Thus, the excess NO_3^- contributes to the loss of XCa..Na by leaching before plants reestablish and actively absorb the anions formed during the flush period. Freezing followed by thawing also stimulates mineralization.

After flooding of aerobic soils, the anaerobic organisms take over and mineralization may be little affected. Soils that tend to mineralize a large amount of N when aerated will tend to mineralize a fairly large amount of N when flooded. Thus, rice farmers commonly apply crop residues and animal manures before flooding to increase the amount of available N for cropping.

A wide variety of organisms participate in N mineralization resulting in low sensitivity of mineralization to soil pH. In some strongly acid situations, however, organic matter accumulates as a mat on the soil surface, owing to a lack of decomposition or mineralization. In bogs where the water is stagnant and N mineralization is minimal, some plants, like the Venus fly trap, snare insects to get N.

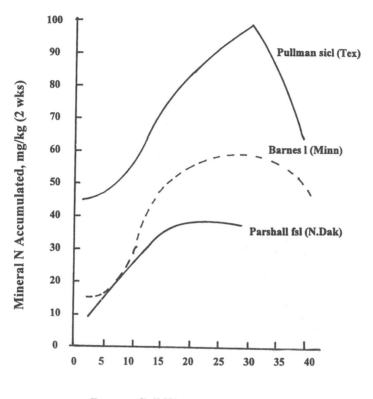

Figure 6.6 Mineral nitrogen accumulation in three soils of varying texture as a function of soil water content. (Redrawn from Stanford and Epstein, 1974. Used by permission of Soil Science Society of America.)

6.3.2   Carbon to Nitrogen Ratio

Mineralization is carried out by a wide variety of heterotrophic organisms that have a dietary need for both C and N. The net amount of N produced by mineralization represents an amount of N in the substrate that is in excess of microbial needs.

The C:N ratio (%C/%N) of the substrate provides an indication of the adequacy of N for the mineralizers and the amount of N that will appear as excess. Humus or soil organic matter (SOM) has a C:N ratio typically in the range of 10 to 12:1, and when it is mineralized, it provides N in excess of microbial needs. This excess accumulates and can be used by roots. During decomposition or rotting, if there is a deficiency of N in the substrate for the microorganisms, they will use whatever available N is present in the immediate environment. Under these conditions, mineralization of N occurs without any N accumulating for uptake by roots. Any plants growing will experience N stress and retarded growth as illustrated in Figure 6.7.

Figure 6.7 Effect of increasing amount of paper mill sludge on growth of corn seedlings. Soil containing one-fourth and one-half of the mix as sludge produced much smaller seedling having nitrogen deficiency symptoms. (From *Fundamentals of Soil Science*, 7th ed. Foth, H. D. © 1984. Reprinted by permission of John Wiley & Sons.)

During the decomposition of substrates, there is a continual loss of C as respiratory CO_2, with an accompanying reuse of mineralized N, resulting in an increase in the N percentage and a decrease in the C:N ratio (Figure 6.8). In time, the microbes that had earlier depleted the soil of any excess available N die, and they become substrate and their N mineralized. Microbes have a relatively high N content, low C:N ratio, and with time the temporary N stress period shown in Figure 6.7 disappears.

Residues of legume crops and farm manures provide an amount of N in excess of mineralization needs, thus creating a surplus that can be used by growing crops. As the C:N ratio of substrates increases, there is relatively less N for the C, and substrates with ratios above 35 are not likely to contain enough N to meet microbial needs. Many materials added to soils, such as straw and sawdust, have C:N ratios of about 80 and 400, respectively. Their addition to soils before planting can create a period of N starvation for crops. It requires about 20 pounds of N per ton of straw to supply the mineralization N deficit of the straw and prevent a period of net immobilization of N.

Virgil and Kissel in 1991 reported on experiments conducted with a variety of mineral soils and substrates for periods longer than 100 days and concluded that the percentage of N mineralized in crop residue could be estimated by:

$$Y = 58.89 - 1.41(C/N) \qquad (6.1)$$

Accordingly, they found that the C/N ratio where zero net N mineralization and immobilization from decomposition of crop residue was about 40. This corresponds to about 1% N in the crop residue.

6.3.3 Mineralization Under Anaerobic Conditions

Anaerobic bacterial degradation of organic matter in submerged soils is characterized by incomplete decomposition with a consequent low yield of energy for the decomposers. The low energy yield results in only 2 to 5% percent of the C in substrates being metabolized into microbial tissue as compared to 30 to 40% assimilated in an aerobic system. This results in a low demand for N for the mineralizing organisms and more N in excess for plant use from the mineralization of a given amount of organic matter. In lowland rice culture, anaerobic decomposition of crop residues and manures results in more mineralized N for crops than in well-drained soils from the decomposition of comparable residues and manures. Thus, organic materials with C/N ratios in excess of 35 or 40 can be incorporated into flooded rice fields with little danger of N stress for the rice crop.

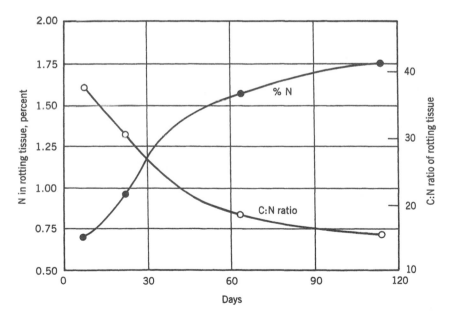

Figure 6.8 Changes in nitrogen content and carbon-nitrogen ratio during the decomposition of barley straw. (From *Introduction to Soil Microbiology*, 2nd ed., Alexander, M. © 1977. Reprinted by permission of John Wiley & Sons.)

6.3.4 Effects of Accessibility and Recalcitrance

Litter on the soil surface, compared to incorporation within the soil, is less subject to mineralization because conditions are less favorable for microbial activity. Within the soil, finely divided organic matter mineralizes faster than coarse particle material. Animals enhance mineralization by comminuting organic matter and moving organic matter to locations more desirable for mineralization.

Mineralization is dependent on enzyme activity. The fact that there is a strong positive correlation between clay content of soils and SOM content suggests that clays may inactivate the proteolytic enzymes in some way. In addition, organic matter may be adsorbed onto clays and made inaccessible for attack by enzymes. Clay-protected decomposable organic matter decomposes about 1% as rapidly as that which is unprotected.

Some organic compounds are recalcitrant (i.e., they are chemically very resistant to breakdown). Water-soluble components disappear in the soil very rapidly when conditions for mineralization are favorable. On the other hand, the most resistant compounds persist in soils hundreds and thousands of years, resulting in a factor of about one million between the mineralization rate of the most readily decomposed and of the most resistant to decomposition. The net amount of N mineralized in a soil is the sum total of the contributions from all the SOM fractions. The contribution of various organic matter fractions to mineralized N in a Mollisol (Chernozemic soil), during a 12-week incubation period is shown in Table 6.3. The 4 to 6% of the N in decomposable biomass and the 6 to 10% of the N in active decomposable non-biomass materials produced 30 and 34%, respectively, of the total amount of N mineralized. Note that 50% of the soil's N considered as old, chemically recalcitrant, contributed only 1% to the mineralized N during the incubation period. For many soils, the average turnover time of SOM is 20 to 40 years; however, the mean residence time of SOM is about 175 years.

Table 6.3 Nitrogen Content and Contributions to Nitrogen Mineralization of Organic Matter Components in a Chernozemic Soil

Component	Percent of soil N	Decomposition, half-life, years	Relative contribution to mineralized N, %*
Biomass	4–6	0.5	30
Active non-biomass	6–10	1.5	34
Stabilized	36	22	35
Old	50	600	1

* During 12 week incubation.
Source: Data from Paul, 1984.

6.4 NITRIFICATION

The starting point for nitrification is the NH_4^+ from mineralization. In aerobic soils with a pH of 6.0 or higher, NH_4^+ is rapidly oxidized by specialized chemoautotrophic bacteria, which are widespread in nature. The process is *nitri-fication*, and occurs in two steps.

Bacteria of the genus *Nitrosomonas* and several other bacteria oxidize the NH_4^+ to nitrite as follows:

$$2NH_4^+ + 3O_2 = 2NO_2^- + 2H_2O + 4H^+ \qquad (6.2)$$

The valence of N goes from -3 to $+3$ and the energy released is sufficient to enable bacteria to fix all the C they need from CO_2. Notice that the first step in nitrification produces protons and is a natural soil acidification process. Nitrifi-cation may significantly lower soil pH when large amounts of ammonium fertil-izers are applied. Little if any nitrite accumulates because the nitrite excreted by *Nitrosomonas* is quickly picked up by bacteria of another genus, *Nitrobactor*, and oxidized to nitrate as follows:

$$2NO_2^- + O_2 = 2NO_3^- \qquad (6.3)$$

The valence of N goes from $+3$ to $+5$. Nitrate is stable in the soil solution and readily moves to roots via mass flow. In the absence of uptake by roots or microorganisms, nitrate is subject to loss from soils by leaching.

Some heterotrophic organisms have been isolated that are nitrifiers. They may be important in some habitats. There is insufficient data to properly assess their importance in soils.

6.4.1   Factors Affecting Nitrification

Ammonification or N mineralization is quite insensitive to the soil environ-ment, and the production of NH_4^+ in soils is quite ubiquitous. Nitrification is much more environmentally dependent. Whether the NH_4^+ accumulates or is nitrified depends on temperature, acidity, and the water and O_2 supply. Little nitrification occurs in wet and cold soils.

The nitrifiers are sensitive to H^+. Their activity is reduced below pH 6.0 and becomes negligible below 5.0. Some soils with pH 4.0 or less, however, may contain some NO_3^- and it appears that the organisms derived from acid soils are frequently more tolerant of H^+. Optimum pH is 6.6 to 8.0 or higher.

All nitrifiers need O_2 and nitrification ceases in its absence. For this reason, nitrification is sensitive to soil structure and water content. Generally in aerobic soils, optimum water content is 50 to 67% of the water-holding capacity. Oxygen diffuses very slowly through water so that nitrification may be occurring in the outer part of an aggregate at the same time that there is denitrification in the interior. Waterlogging or flooding suppresses nitrification, and the NH_4^+ produced in rice paddies tends to be absorbed by roots and not be nitrified.

Nitrification is temperature sensitive and occurs mainly in the range 5 to 40°C with an optimum between 30 and 35°C.

6.4.2 Summary Statement

Nitrification occurs rapidly in most well-drained and moist agricultural soils with a pH 6.0 or higher. Nitrate is the main form of N absorbed by roots because it is the form most readily available in soils. In acid forest soils, nitrification is commonly inhibited by acidity, and it is believed that plants in forests absorb much of the N as NH_4^+. Under the anaerobic conditions of flooded soils, the lack of O_2 inhibits nitrification, and NH_4^+ is the predominant form of N absorbed. Nitrification in the absence of plant uptake of nitrate results in nitrate accumulation in soils, in the absence of denitrification or leaching.

6.5 IMMOBILIZATION

After the uptake of NH_4^+ and NO_3^- by heterotrophic microorganisms and roots, the conversion of the mineral N into organic N is called *immobilization* (see Figure 6.1). Mineralization and immobilization are two opposing and interdependent processes involved in the decomposition and formation of organic material. The continuous transfer of mineralized N into organic products of synthesis and of immobilized N back into inorganic decay products has been called the mineralization-immobilization turnover, MIT. The MIT strongly affects the supply of available nitrogen in soils.

Nitrification can occur in between mineralization and immobilization, controlling whether the available form of N will be NH_4^+ or NO_3^-. The available N is absorbed by both the heterotrophs and plant roots; thus, both are benefited. Moreover, root exudates and sloughed-off root cells provide a readily available source of energy and C for the heterotrophs. The heterotrophs are therefore additionally benefited by the plants. Soil conditions favorable for growth of the heterotrophs are favorable for plant growth, which means that there is a strong correlation between mineralization and plant immobilization of N. As a consequence, mineralization and immobilization comprise a subcyle of the soil's N cycle, one that governs the level of available N in unfertilized soils at any given instant. Since mineralization and immobilization occur simultaneously, plants may have an adequate supply of available N, even though soil tests indicate that little N is available at a given instant.

6.6 DENITRIFICATION

Denitrification is the chemical reduction of nitrate and nitrite to gaseous forms, nitric oxide, nitrous oxide, and dinitrogen:

$$NO_3^- \ @ \ NO_2^- \ @ \ NO \ @ \ N_2O \ @ \ N_2$$

nitrate	nitrite	nitric oxide	nitrous oxide	dinitrogen	
(5)	(3)	(2)	(1)	(0)	(6.4)

The valence of N decreases from 5 or 3 to 2, 1, or 0. Products formed and utilized during denitrification in the sequence from nitrate to N_2 are shown in Figure 6.9 for a soil having a large amount of nitrate due to pretreatment.

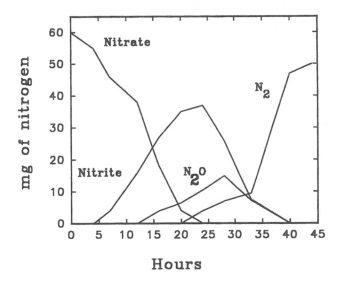

Figure 6.9 Products produced during denitrification in loam soil with pH 7.8. (From Cooper and Smith, 1963. Used by permission of the Soil Science Society of America.)

The capacity for true denitrification, formation of N_2, is restricted to certain chemoheterotrophic bacteria. These bacteria are aerobes; however, under anaerobic conditions, they use NO_3^- for their electron acceptor instead of O_2. They need a decomposable source of organic matter to supply C and electrons. Thus, the two most important conditions needed for denitrification are an anaerobic environment and a C source.

6.6.1 Factors Affecting Denitrification

Because a lack of O_2 favors the chemical reduction of nitrate and nitrite to gaseous forms, denitrification is affected by soil structure and water content, which is the converse of nitrification. Anaerobic microenvironments are created in soils when the O_2 demand exceeds the supply. A significant amount of denitrification and loss of gaseous N may occur in many well-drained soils through denitrification in the interiors of aggregates. Flooding creates conditions very conducive to denitrification, if nitrate is present. Since NO_3^- is produced in an aerobic environment and denitrification occurs in an anaerobic environment, denitrification is enhanced by the alternate wetting and drying of soils.

Acidity affects both the rate of denitrification and the type of gas produced. Denitrifiers are sensitive to H^+, although denitrification has been reported to be rapid in a soil with a pH of 4.7. Above pH 6.0, the major gas produced is N_2, and in more acid environments the liberation of N_2O becomes pronounced.

Nitrification is very slow at 2°C and increases to a maximum at 25°C or higher. Significant denitrification can occur when it is too cold for crops to be growing. During the late fall and early spring in the temperate regions, there is a substantial loss of N through denitrification.

The growth of plant roots may inhibit denitrification by serving as a sink for NO_3^- and creating drier soil with a greater O_2 diffusion rate. On the other hand, root exudates provide a source of carbon and deplete O_2 near their surfaces to stimulate denitrification. When NO_3^- moves by leaching below the zone of biological activity, or C source, it is beyond the sphere where denitrification occurs; it may remain unchanged and move farther downward and pollute the groundwater. In agriculture, denitrification represents a large loss of nitrogen, and there are few practical things that can be done to prevent it.

6.6.2   Denitrification in Wetland Rice Soils

Wetland rice soils, paddy soils, have aerobic and anaerobic zones. Water on top of the soil contains O_2 as a result of dissolution of atmospheric oxygen into the water. Also, oxygen is released into the water by photosynthesizing plants. Oxygen from the overlying water layer diffuses a short distance into the upper few centimeters of soil creating a thin layer of aerobic soil. Underneath is the major root zone, which is water saturated and anaerobic. These layers are shown in Figure 6.10.

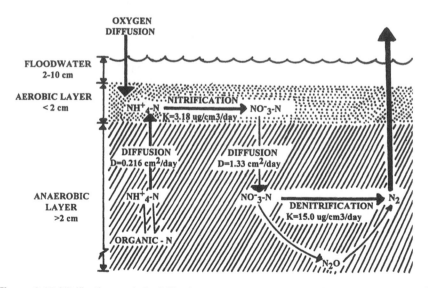

Figure 6.10 Nitrification and denitrification sequence reactions in flooded soil. (Adapted from Patrick and Reddy, 1977.)

Organic N is mineralized within the anaerobic layer and ammonium is formed. The NH_4^+ diffuses upward from the anaerobic water-saturated layer into the thin aerobic soil layer where nitrification occurs. Subsequent diffusion of

NO_3^- downward into the anaerobic layer results in denitrification and loss of gaseous N_2 (see Figure 6.10).

Drying of water-saturated soil during the growing season results in much nitrification. Subsequent flooding results in denitrification. Large N losses from denitrification occur in rain-fed systems because fields alternately dry and resaturate in the absence of available irrigation water. The shifts in inorganic N in a flooded rice field during the cropping season are shown in Figure 6.11. Early in the season, nitrate is the dominate form of N that accumulates in aerobic soil before flooding. Flooding results in anaerobic soil and the loss of nitrate N by denitrification and also some loss by leaching. Under the anaerobic conditions of flooding, ammonium N is produced and accumulates as the dominant form of N taken up by the rice. After water is withdrawn and aerobic conditions are restored for harvesting, nitrate accumulates again as the dominant form of inorganic N in the soil. Frequently, a post-rice legume crop will be grown that utilizes the accumulating nitrate. The legume crop produces food grain and provides N-rich residues that will mineralize and increase available N for the next rice crop.

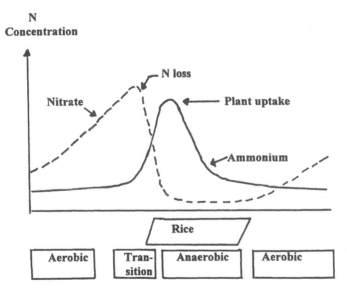

Figure 6.11 Inorganic nitrogen changes and plant uptake during a period of rice production. (Redrawn from Buresch and DeDatta, 1991. Used by permission of the American Society of Agronomy.)

6.7 AMMONIUM FIXATION

The ammonium formed by mineralization appears in the soil solution and establishes an equilibrium with the exchangeable ammonium. The similarity in ionic radius and energy of hydration of NH_4^+ and K^+ causes the ions to compete for specific adsorption and fixation in micaceous minerals. Soil drying enhances fixation in 2:1 expanding minerals. The more fixed K a soil contains, the less ammonium it will fix. The upper limit of ammonium fixation in most field soils is about 1 to 2 cmol kg^{-1}.

About 10% of the N in soils may be fixed, and its distribution in the soil profile parallels that of the clay. Just as with K$^+$, an equilibrium for NH$_4^+$ exists between the ions that are exchangeable and those that are in solution, and between the ions that are exchangeable and those that are fixed. Ammonium fixation is more rapid than release and the fixed ammonium appears to be a slow-release N reservoir for nitrifying bacteria and roots. Although considerable fixed NH$_4^+$ may exist in soils, it is of minor importance in meeting the daily N needs of growing plants. The agronomic significance of ammonium fixation is clouded by conflicting experimental results.

6.8 SUMMARY STATEMENT FOR SOIL NITROGEN CYCLE

Nitrogen is continually added to soils by fixation, precipitation, and gaseous adsorption. The N content of the soil, however, does not increase continually but tends to reach an equilibrium or steady-state where additions are balanced by losses. The losses or leaks from the cycle are due to volatilization, leaching, and denitrification. As a result of these losses, there is a need for a constant influx of N by fixation to maintain a steady N content.

Soils do not become so enriched with N (except maybe some Histosols) that they produce high grain yields for a long period of time without N fertilizer. Rather, the soil is characterized by a slow and consistent mineralization-immobilization turnover of N without extended periods of a naturally high level of available N. Most of the time, soil microorganisms are quite inactive due to a lack of decomposable substrates. Thus, there is commonly a need for frequent additions of N as fertilizer to produce high yields of nonleguminous crops. Excessive N fertilization, however, can exceed the immobilization capacity and result in unused nitrate leaching to the water table.

6.9 NITROGEN UPTAKE BY PLANTS

Both NH$_4^+$ and NO$_3^-$ are commonly present in soil solutions, and both are readily taken up by roots. The rapid nitrification rates in well-aerated soils with pH 6 to 8 means, however, that plants absorb predominately NO$_3^-$. Nitrate remains soluble in the soil solution and is readily moved to plant roots by mass flow. It has been estimated for corn that 79, 20, and 1% of the nitrogen at root surfaces are there through mass flow, diffusion, and root interception, respectively.

In some soils, acidity and anoxia greatly inhibit nitrification, and the predominate form of available N is NH$_4^+$. In general, calcifuge plants that naturally grow under acid conditions prefer NH$_4^+$ and calcioles, plants with a wide pH tolerance prefer NO$_3^-$. At the extreme are cranberry plants, which are unable to absorb or metabolize NO$_3^-$. Rice grown on flooded soil, where nitrification tends to be inhibited, is more productive with ammonium than with nitrate fertilizer. It has not been proved, however, that NH_4^+ is clearly the most efficient form of N for rice.

When plants absorb N mainly as NH_4^+, there is an excess of cation uptake compared to anion uptake, and H^+ is excreted, which decreases the pH in the rhizosphere. When uptake is mainly NO_3^-, there is an excess uptake of anions, and OH^- is excreted, which increases the pH of the rhizosphere. In a respiring rhizosphere, the hydroxyl reacts with carbon dioxide to form HCO_3^-. The question of whether NO_3^- or NH_4^+ is the superior or most efficient form of N for plants is still unanswered.

6.9.1 Enhanced Ammonium Nutrition

Nitrification inhibitors have been used to delay the conversion of ammonium to nitrate as a means to reduce nitrate leaching. The development of more efficient nitrification inhibitors is increasing our ability to control the form of N available for plant uptake. This increased ability, and the theory that plants utilize ammonium N more efficiently than nitrate, has resulted in the development of the concept of enhanced ammonium nutrition, EAN. Enhanced ammonium nutrition generally means an increased uptake of ammonium relative to nitrate. Enhanced ammonium nutrition is being studied as a means to increase crop yields and reduce nitrate pollution of groundwater. Current cultivars have been selected for performance with predominantly nitrate N. Plant breeding programs are being developed to exploit the potential benefits of EAN.

6.10 NITROGEN DISTRIBUTION AND CYCLING IN ECOSYSTEMS

The major pools of organic matter in ecosystems are (1) the living and dead standing vegetation, (2) organic residues at and near the soil surface, and (3) organic matter within the soil. About half of the organic matter in a forest exists above ground and about half below ground. Grasslands, by contrast, generally have over 90% of the organic matter within the soil. Because soil organic matter has a lower C/N ratio than the organic materials above the soil, the major pool of N in ecosystems is SOM.

6.10.1 Nitrogen and Carbon Pools in Grassland Ecosystems

Roots are the most important source of organic matter in grassland ecosystems. A mixed-grass prairie in North Dakota produces about 1.4 tons per acre of above ground growth and about 3.6 tons per acre of growth below ground as roots. Thus, it is not surprising that a grassland in summer may have about 5% of the organic matter in vegetation, 5% as litter, and 90% of the organic matter within the soil. As with forests, SOM has a lower C/N ratio, as compared to vegetation and litter, resulting in the SOM as being the overwhelming reservoir of N in grasslands. About 98% of the N in a grassland ecosystem in Saskatchewan, Canada, was reported as existing within the soil.

6.10.2 Nitrogen and Carbon Pools in Forest Ecosystems

The percentages of organic C in a 49-year-old Ponderosa pine forest ecosystem in Arizona were 38, 9, and 53% for the standing crop, forest floor, and soil, respectively. By contrast, the distribution of N was about 5, 5, and 90% for vegetation, forest floor, and soil, respectively, as shown in Table 6.4. Note that the soil N includes root N, which contain less than 2% of total N.

Table 6.4 Quantity and Distribution of Carbon and Nitrogen in a Ponderosa Pine Ecosystem

Components	Carbon kg/ha	Carbon %	Nitrogen kg/ha	Nitrogen %	C:N ratio
Standing crop	74,723	37.7	322	5.4	232
Forest floor	17,079	8.6	291	4.9	59
Below ground roots	17,000	8.6	93	1.6	183
Soil organic matter	89,300	45.1	5,200	88.1	17
Total	198,102	100	5,906	100	

Source: Data from Klemmedson, 1975.

The two most important sinks for supplying available N to the plants in the ecosystem are the soil and the forest floor. These components are accessible for decomposition and have the lowest C/N ratios. The large C/N ratio of standing vegetation means that the production of wood places a meager N demand on the soil compared to the production of grain crops. Less than 10 kg/ha of N accumulated annually in the standing crop or vegetation during a 49-year period.

6.10.3 Nitrogen Cycling in a Natural Ecosystem

Bormann and associates made a detailed study of the N budget for a 13.2 ha forested watershed in New Hampshire where the dominant trees were sugar maple, American beech, and yellow birch. The forest was about 55 years old and was aggrading, with a net annual accretion of biomass. The annual addition of N was equal to about 21 kg/ha with 14 kg due to N fixation and 7 kg contained in the bulk precipitation (Figure 6.12). Of the N added to the ecosystem annually, the hydrologic loss is only 4 kg/ha and 17 kg/ha are accreted to long-term storage within the ecosystem.

The annual total utilization of N by plants is 120 kg/ha. This large annual N use is accomplished, in large part, by the recycling of N. Each year about 40 kg/ha of N is withdrawn from tree storage in the spring, used during the summer, and replaced in plant storage before leaf senescence in the fall. A considerable amount, 63 kg/ha, is returned to the litter layer and soil by canopy wash and litter fall. On a long-term basis, the uptake of 80 kg/ha of N is accounted for with 17 kg from long-term accretion and 63 kg returned from canopy wash and litter fall. Most of the N taken up, however, was due to the net mineralization of organic

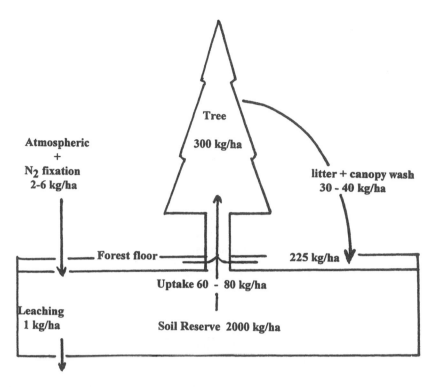

Figure 6.12 Nitrogen cycling in a forest ecosystem. (Redrawn from W.L. Pritchett and R. F. Fisher, 1979. Used by permission of John Wiley & Sons.)

N within the soil and litter layer. With a net annual input of only 17 kg/ha, the forest was able to utilize 120 kg/ha of N in growth.

6.10.4 Nitrogen Pools and Cycling in Agricultural Soils

Mineral soil parent materials are low in organic matter. During soil formation, the organic matter content increases with time to a quasi-equilibrium where additions are balanced by losses. Under these conditions, changes in SOM and soil N occur very slowly and the soil has a near steady-state N content. When such soils are converted to agricultural land, the C and N sinks, such as standing trees and forest floor, are destroyed. All of the N and C can be considered to exist as components in SOM, except for some crop residues and microbial biomass. With conversion to agricultural use, there is a rapid loss of organic matter and N, so that in a hundred years, a new quasi-equilibrium N content is established with an organic C and N content about half of the original as shown in Figure 6.13.

The figure is based on an organic matter turnover model and projects changes with time in three pools of SOM as a result of the conversion of a grassland into a cropland. These pools of organic matter also represent the three pools of N in the soil. One pool consists of plant residues and other active fraction materials

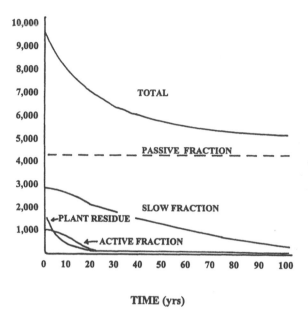

TIME (yrs)

Figure 6.13 Organic matter changes in conversion of grassland to cropland. (From Anderson and Coleman, 1985. Used by permission of the Soil and Water Conservation Society of America.)

such as microbial biomass and products. These are decomposable materials and there is a very rapid projected decrease with time. There is a projected gradual decline in the slow fraction, slowly decomposing organic matter consisting mainly of ligniferous products. By contrast, there is little if any projected change with time of the passive fraction consisting of chemically stabilized materials. One can conclude from the model that the organic matter content and N content of old agricultural soils is determined mainly by the amount of passive organic matter.

The decay rates of various kinds of organic matter components are given in Table 6.5. They are grouped into three pools representing the pools shown in Figure 6.12, with the addition of clay- or mineral-protected fractions. Plant residues, roots, decomposable unprotected organic matter and soil organisms have a decay rate of about 3 to 8% per day. These materials comprise the labile or active pool and have a turnover time of a few days. The passive pool consists of decomposable organic matter protected by clays and has a decay rate of about 0.08% per day. The slow pool consists of recalcitrant organic matter with decay rates ranging from 0.0008 to 0.000008% per day, depending on whether protected or not. Turnover time for the passive and slow pools is of the order of decades and hundreds of years, respectively. For our consideration, the passive and slow pools make up the stable soil N pool.

Table 6.5 Content and Decay Rates of Organic Matter Fractions in a Canadian Grassland Soil

Fraction	Decay rate per day, %	Percent of total
Active	8.0	1.2
Plant residue	8.0	9.0
Soil organisms	8.0	4.0
Decomposable, unprotected	3.0	1.3
Passive decomposable, protected	0.08	54
Slow		
Recalcitrant, unprotected	0.0008	15
Recalcitrant, protected	0.000008	15

Source: Data from Van Veen and Paul, 1981.

In the production of annual crops there is no pool of N held within living plants that can be utilized and carried over from one year to another. On the contrary, the nonleguminous crops must depend entirely on the soil for their N needs. Unless the soil is supplied with labile or readily decomposable organic matter with narrow C/N ratio, an old agricultural soil has very limited capacity to supply available N for cropping. At the end of the season, most of the N in the crop is harvested and removed from the field. Thus, the amount of N recycled annually is dependent on a small amount of N in crop residues because any mineralized N from the passive and slow pools will tend to be offset by an equivalent amount of N immobilized into "new" passive and slow organic matter. Without the addition of N in manures and fertilizers, the long-time yield of grain crops declines to a stable plateau, reflecting the net N accretion of N from natural additions as N fixation and precipitation. This amount of N can produce about 8 to 10 bushels of grain per acre annually.

6.10.5 Summary Statement

The constant nature of the N content of the passive and slow fractions, from year to year, means that their contribution to mineralized N is offset by an equal amount of immobilized N each growing season. Therefore, a net mineralization of N in excess of immobilization is highly dependent on the mineralization of N from the active fraction. The soil appears designed to maintain a low but fairly constant level of available N, rather than a high level of available N during the growing season for the production of high yielding crops. Any significant buildup of available N, if it occurs, is subject to loss by denitrification or leaching. This is in stark contrast to many soils that can provide high available levels of certain nutrients for long periods of time. Historically, farmers have increased crop yields by amending soils with decomposable materials having low C/N ratio, such as animal manures and crop residues, to create a temporary period of high N availability for crop production. Today, nitrogen fertilizers are used for the production of virtually all nonleguminous commercial crops.

6.11 BASIS OF NITROGEN FERTILIZER RECOMMENDATIONS

Variations in mineralization, denitrification, and leaching rates from one soil to another and from one year to another complicate the use of a soil test for making N fertilizer recommendations. The recommendations for N fertilizers are based primarily on yield response data obtained from N fertilizer rate experiments.

6.11.1 Grain Crops

Fertilizer rate experiments in Illinois were used to develop Table 6.6, which give economically optimum corn yields and optimum N rates per bushel of corn produced. The economically optimum N fertilizer rate for corn is based on a corn/nitrogen price ratio: the price of a bushel of corn divided by the price of a pound of fertilizer N. Corn grown at Brownstown had the least ability to absorb N fertilizer and profitably increase yields; the economic optimum yield was 83 bushels/acre with a corn:nitrogen ratio of 10. Soils at Urbana were the best for corn production; the economically optimum yield was 171 bushels/acre with a corn:nitrogen ratio of 10. Notice in Table 6.6 that it took less N per bushel of yield to produce the optimum yield at Urbana as compared to Brownstown: 1.17 versus 1.3. Based on the optimum N rate, pounds of N per bushel of yield multiplied by the optimum economic yield, the amount of N fertilizer recommended for Brownstown is 108 (83 × 1.3) pounds of N per acre for a corn:nitrogen ratio of 10. For Urbana, the N fertilizer recommendation is 200 (171 × 1.17) pounds/acre of N. The basic equation for the N fertilizer recommendation is:

$$\text{pounds N per acre} = YG \times \text{optimum N rate} \tag{6.5}$$

where YG equals yield goal (bushels/acre) and the optimum N rate is the pounds of N needed per bushel of yield to produce the economically optimum yield. Some examples of N fertilizer recommendation equations for corn and sorghum production in Iowa are given in Table 6.7.

Table 6.6 Economic Optimum Yield* and Optimum Nitrogen Rate Experimentally Determined for Continuous Corn at Four Locations in Illinois

| | Corn:nitrogen price ratio | | | |
| | 10:1 | | 20:1 | |
Location	Yield	Rate	Yield	Rate
Brownstown	83	1.30	86	1.47
Carthage	144	1.22	147	1.29
DeKalb	141	1.28	143	1.31
Urbana	171	1.17	173	1.24

* Yield in bushels per acre and optimum N rate in pounds of nitrogen per bushel.

Source: Data from *Illinois Agronomy Handbook*, 1986, University of Illinois Coop. Ext. Ser., Urbana, IL.

Table 6.7 Nitrogen Recommendation Procedures for Corn and Sorghum Based on Yield Goal and Optimum Nitrogen Rate per Bushel in Iowa

Soil	N equation for pounds of N association per acre
Moody	N = YG × 0.9
Marshall	N = YG × 1.1
Clarion-Nicollet-Webster	N = YG × 1.2
Tama-Muscatine	N = YG [180 + (YG − 150)] × 1.3

Source: Data from *General Guide for Fertilizer Recommendations in Iowa*, 1982, Iowa State Univ. Coop. Ext. Ser., Ames, IA.

The basic N fertilizer recommendation for corn is modified by consideration of the previous crop and the amount of manure to be applied. The equation used in Michigan for corn N recommendations as pounds per acre is:

$$N = \left[(YG \times 1.36) - 27\right] - \left[40 + (0.6 \times \% \text{ ls})\right] - (4 \times \text{fm}) \qquad (6.6)$$

The ls is for legume stand, which is evaluated in terms of number of alfalfa or clover plants per unit area. If an alfalfa stand was rated very good, 100%, then 100 pounds ($40 + 0.6 \times 100$) would be subtracted to account for the N contribution of the previous alfalfa crop. The fm refers to tons of farm manure applied per acre. A 10 ton/acre application of manure would mean a subtraction of 40 (4×10). For a YG of 150 bushels/acre, when corn follows an excellent stand of alfalfa or clover and 10 tons/acre of manure are applied, the N recommendation is

$$\text{pounds } N = 177 - 100 - 40 = 37 \qquad (6.7)$$

Sometimes a previous soybean crop is equated to 40 pounds of fertilizer N. This basic method is also used to make N recommendations for small grains and other nonlegume crops. The method can be called the N balance method in that an attempt is made to determine crop N needs and the amount of N expected to be available. The difference between need and the predicted availability is the fertilizer recommendation.

Other factors are used to modify the recommendation, depending on the crops and conditions. In areas of limited precipitation and leaching, there is opportunity for NO_3^- to be carried over. A test to determine the amount of NO_3^- in the root zone is used in many of the western states. The samples for nitrate should be obtained to a minimum of 60 cm and, if possible, to at least 120 cm. Montana recommends an adjustment when straw with a high C:N ratio is plowed down before planting. Twenty pounds/acre (22 kg/ha) of N is recommended to account for the immobilization of N by microorganisms. Arizona makes an adjustment for N in irrigation water. A summary of some N fertilizer evaluation systems used in various regions of the U.S. is given in Table 6.8. The pre-sidedress nitrate test has been developed in the northeastern states. Environmental concerns dictate that we should be as precise as possible in determining the correct quantity of nitrogen fertilizer to apply and as precise as possible in timing the application.

Table 6.8 Summary of Some Factors Used in Current Nitrogen
Evaluation Systems in the U.S.

Region	Crop	Average nitrogen factor*	Average nitrogen credits**		
			Manure	Soybeans	Alfalfa
Northeast	corn	1.16	4.5	NA	134
Mid-Atlantic	corn	1.18	6.0	16	71
Southeast	corn	1.25	NA	27	49
Midwest	corn	1.24	4.8	29	89
West corn	corn	1.41	4.8	32	80
West	wheat	2.10	—	—	—
Southwest	wheat	2.15	NA	NA	80
Northwest	wheat	2.40	NA	NA	NA

* Pounds of nitrogen per bushel of yield goal.
**Pounds per ton or pounds of nitrogen per acre. NA refers to inadequate
data or may mean that in the region the material is an infrequent source
of N.
Source: Selected data from Meisinger, 1984.

Two known factors form a basis for a management system that can help minimize
loss of nitrate through leaching. First, the demand for nitrogen by corn is very
high during a period of growth which accelerates after the corn is 30 cm high
(12 inches). And secondly, sidedress nitrogen applications are often more efficient
at furnishing fertilizer nitrogen for corn. A pre-sidedress nitrate test (the test has
been referred to by a number of names: Magdoff pre-sidedress nitrate test, pre-
sidedress nitrate test, late-spring nitrate test, mid-June nitrate test, etc.) has been
developed to determine soils that do not need nitrogen application as a sidedress
treatment for maximum yield. Rapid methods of nitrate analysis have greatly
improved the utility of this test. These include a nitrate electrode, nitrate quick
test, and automated flow analysis. A rapid turn-around time allows for application
of sidedressed nitrogen after the results of the test are known. Considerable
research has shown that corn growing on a soil containing greater than 20 to 30
mg N/Kg soil in the surface 30 cm (one foot) as nitrate when the corn is 15 to
30 cm high will not respond to further additions of nitrogen fertilizer. The test
is successful in defining the nitrogen status of soils that have received manure
and other organic residues. It should also be noted that in a north-central regional
study conducted in 1989 and 1990, there was no response to applied nitrogen at
sidedress time if corn followed alfalfa even if the pre-sidedress nitrate test showed
less than 20 mg N/kg soil in the top 30 cm.

6.11.2 Legume Crops

Legume crops vary in their N-fixing efficiency. Alfalfa and clover are not
benefited by N fertilizer, except perhaps to become more rapidly established when
the N-fixing system is not fully operative. State agricultural experiment stations
are becoming more conservative in recommending N to avoid possible groundwater contamination, and they are eliminating N recommendations for soybeans.

In addition, the application of N fertilizer to efficient N-fixing legumes may reduce N fixation without affecting yields (see Figure 6.14). On the other hand, some of the least efficient N-fixing legumes, such as peanuts in Georgia, peas in Washington, and navy beans in Michigan, are routinely fertilized with N.

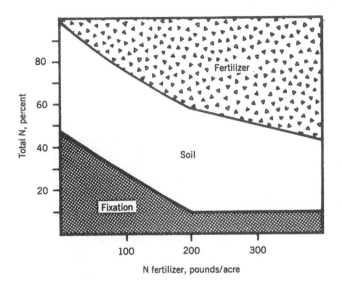

Figure 6.14 Percent of total nitrogen in soybeans derived from fertilizer, soil, and symbiotic fixation. (Data from Johnson, Welch, and Kurtz, 1974.)

6.11.3 Crops Grown on Organic Soils

Organic soils, Histosols, contain a greater quantity of total N than mineral soils. As a consequence, the N mineralization potential is very high and many crops are grown without N fertilizer. These soils require drainage and tend to be wet and cold in the spring causing a slow mineralization rate. During this season, some crops respond to small applications of N fertilizer.

REFERENCES

Alexander, M. 1977. *Introduction to Soil Microbiology.* 2nd ed. John Wiley & Sons, New York.

Anderson, D. W. and D. C. Coleman. 1985. The Dynamics of Organic Matter in Grassland Soils. *J. Soil and Water Con.* March-April, pp. 211–216.

Baethgen, W. E. and M. M. Alley. 1986. Nonexchangeable Ammonium Nitrogen Contribution to Plant Available Nitrogen. *Soil Sci. Soc. Am. J.* 51:110–115.

Bormann, R. H., G. E. Likens and J. M. Melillo. 1977. Nitrogen Budget for an Aggrading Northern Hardwood Forest Ecosystem. *Science* 196:981–983.

Buresh, R. J. and S. K. DeDatta. 1991. Nitrogen Dynamics and Management in Rice-Legume Cropping Systems. *Adv. Agron.* 45:1–59.

Cooper, G. S. and R. L. Smith. 1963. Sequences of Products Formed During Denitrification in Some Diverse Western Soils. *Soil Sci. Soc. Am. Proc.* 27:659–662.

Delwiche, C. C. 1969. Nitrogen Fixation. *The Science Teacher.* (J. Nat. Sci. Teachers Assoc.) 36:14–21.

Drury, C. F. and E. G. Beauchamp. 1991. Ammonium Fixation, Release, Nitrification, and Immobilization in High- and Low-Fixing Soils. *Soil Sci. Soc. Am. J.* 55:125–129.

Evans, H. J. and L. Barber. 1977. Biological Nitrogen Fixation for Food and Fiber Production. *Science* 197:332–339.

Foth, H.D. 1984. *Fundamentals of Soil Science.* 7th ed. John Wiley & Sons, New York.

Hauck, R. D., Ed. 1984. *Nitrogen in Crop Production.* Am. Soc. Agron., Madison, WI.

Iowa State Univ. Coop. Ext. Ser. 1982. *General Guide For Fertilizer Recommendations in Iowa.* Agron. Dept., Ames.

Johnson, J. W., L. F. Welch and L. T. Kurtz. 1974. Soybean's Role in Nitrogen Balance. *Illinois Research* 16:6–7.

Keller, G. D. and D. B. Mengel. 1986. Ammonia Volatilization From Nitrogen Fertilizers Surface Applied to No-till Corn. *Soil Sci. Soc. Am. J.* 50:1060–1063.

Klemmedson, J. O. 1975. Nitrogen and Carbon Regimes in an Ecosystem of Young Dense Ponderosa Pine in Arizona. *Forest Sci.* 21:163–168.

Kulasooriya, S. A., P. A. Roger, W. L. Barraquio and I. Watanabe. 1980. Biological Nitrogen Fixation by Epiphytic Microorganisms in Rice Fields. *IRRI Res. Paper Series No. 47.* IRRI, Manila.

Meisinger, L. L. 1984. Evaluating Plant-Available Nitrogen in Soil-Crop Systems. *Nitrogen in Crop Production.* R. D. Hauck, Ed. Am. Soc. Agron., Madison, WI.

Patrick, W. H. and C. N. Reddy. 1978. Chemical Changes in Rice Soils. *Rice and Soils,* pp. 361-379. International Rice Res. Institute, Manila.

Paul, E. A. 1984. Dynamics of Organic Matter in Soils. *Plant and Soil.* 76:275-285.

Phillips, D. A. and T. M. DeJong. 1984. Nitrogen Fixation in Leguminous Plants. *Nitrogen in Crop Production.* R. D. Hauck, Ed. Am. Soc. Agron., Madison, WI.

Pritchett, W. L. and R. F. Fisher. 1979. *Properties and Management of Forest Soils.* 2nd ed. John Wiley & Sons, New York.

Savant, N.K. and S.K. DeDatta. 1982. Nitrogen Transformations in Wetland Rice Soils. *Adv. Agron.* 35:241–302.

Stanford, G. and E. Epstein. 1974. Nitrogen Mineralization-Water Relations in Soils. *Soil Sci. Soc. Am. Proc.* 38:103–107.

Stanley, F. A. and G. E. Smith. 1956. Effect of Soil Moisture and Depth of Application on Retention of Anhydrous Ammonia. *Soil Sci. Soc. Am. Proc.* 20:557–561.

Stevenson. F. J., Ed. 1982. Origin and Distribution of Nitrogen in Soil. In *Nitrogen in Agricultural Soils.* Agronomy 22, Am. Soc. Agron., Madison, WI.

Teyker, R. H., W. L. Pan and J. J. Camberato. 1992. Enhanced Ammonium Nutrition: Effects on Root Growth. In *Roots of Plant Nutrition Proc.,* pp. 116–156. Potash and Phosphate Institute, Atlanta.

Univ. Ill. Coop. Ext. Ser. 1984. *Illinois Agronomy Handbook 1985-1986.* Agron. Dept., Urbana, IL.

Van Veen, J. A. and E. A. Paul. 1981. Organic Carbon Dynamics in Grassland Soils. *Canadian J. Soil Sci.* 61:185–201.

Virgil, M. F. and D. E. Kissel. 1991. Equations for Estimating the Amount of Nitrogen Mineralized from Crop Residues. *Soil Sci. Soc. Am. J.* 55:757–761.

CHAPTER 7

Phosphorus

When scientists began to add nutrients to soils to improve their fertility, phosphorus (P) was soon discovered to be one of the limiting elements. Indeed, P became known as the *master key to agriculture* because lack of available P in soils limited the growth of both cultivated and uncultivated plants. To correct this deficiency, farmers have added P to soils in the form of manures, minerals, or manufactured fertilizers. Now, many soils are sufficient or high in P and the waters that drain from some agricultural lands are also high in P, resulting in increased growth of algae and other plants in surface waters. Erosion of high-P soil has deposited sediments that are high in P into surface waters. Consequently, in areas of the world where farms are managed to obtain maximum yields, environmental concerns are great. Thus, during the past decade, P has become known as the key to eutrophication (defined as surface water becoming highly productive in algae and aquatic weeds). Managing our entire resources, soil, water, and fertilizer, requires a complete understanding of the forms of P in soils, changes which evolve when soluble P is added to soils, and the interactions of P and soil reaction or pH.

7.1 INVENTORY OF THE EARTH'S PHOSPHORUS

For convenience, the earth's supply of P can be divided into P that is sufficiently concentrated to be mined and the P which exists in the soil and other geologic material where the P concentration is too low to be mined but the P becomes slowly soluble for plant or other biological growth. Natural soils will contain from 50 to over 1,000 mg of total P per kilogram of soil. Of this quantity, from 30 to 50% may be in organic form in mineral soils.

Deposits of rock phosphate for mining are relatively common throughout the world. The largest deposits and production in the U.S. are in Florida and North Carolina. Significant deposits are also located in Utah, Wyoming, Idaho, Tennessee and Montana. Very large deposits of rock phosphate exist in other countries, for example, Morocco in Africa, Russia, Brazil in South America, China in Asia and in Australia. Lesser quantities exist in many other countries. The deposits are marine in origin and the rock phosphate mined will be fluoroapatite, chloro-apatite, and hydroxyapatite. Substitutions with carbonate are common. Mining of these deposits forms the basis of the phosphate fertilizer industry.

7.2 PHOSPHORUS CYCLING IN SOIL DEVELOPMENT

Apatite is the principal mineral supplying P prior to extensive soil weathering. Most frequently, apatite is found with calcium carbonate in deposits of *francolite*. Apatites account for more than 95% of the P in igneous rocks and have a general chemical form of:

$$Ca_{10}(X)_2(PO_4)_6$$

where X can be F^-, Cl^-, OH^- or CO_3^-.

7.2.1   Forms in Young and Moderately Weathered Soils

Generally, mineral deposits largely occur as fluoroapatite, although chloroa-patites are quite common. Fluoroapatite is the predominant form in soils, but hydroxyapatite is found in many calcareous soils. Before calcareous soils have been extensively weathered, apatite is their only source of P; the lack of P in solution will limit biological growth. Moreover, in the first stages of the evolution of many soils, the small amount of P available is a severe limitation to plant growth. Under these conditions, the competition for P among microorganisms and macroorganisms is so strong that the P weathering from apatite will be rapidly incorporated into biological tissue and then slowly accumulate in organic com-pounds or humus (SOM). The very low concentration of P in solution limits the opportunity for P to move within the soil. This state for P is depicted by the left side of Figure 7.1. There is little soluble P, apatite is the predominant mineral form, and organic P is slowly accumulating.

How limiting the small amount of available P can be in the early stages of soil evolution is seen by applying solubility product principles to dissolution of apatite. Using, for example, $Ca_{10}(OH)_2(PO_4)_{10}$, which has a $pK_{so} = 114.6$ in equilibrium with CaF_2 with a $pK = 10.41$, we may calculate that total P is less than 1 part per billion (ppb) in solution, assuming the soil contains free $CaCO_3$ in equilibrium with the atmosphere at a pH of 8.3. At a pH of 7, the level would be about 8 ppb. Thus, at pH between 7 and 8.3, apatite has a very low solubility, which together with its crystalline nature accounts for the persistence of apatite in soils that are young or only moderately weathered. The fertility of a soil will increase during this phase of soil formation as organic P accumulates. Although the level of soluble P will be quite low, turnover of organic P as microorganisms decompose SOM furnishes soluble P that can be absorbed by plants in compe-tition with microorganisms.

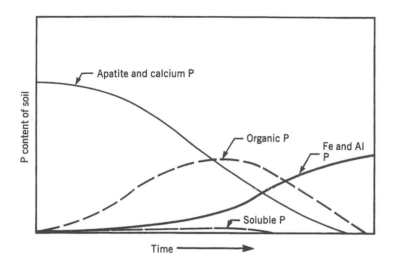

Figure 7.1 The relative distribution of soil phosphate forms as related to time of soil development.

Hydrolysis of soil minerals and leaching during soil evolution remove cations, and soil pH is lowered. The greater acidity at lower pH increases the solubility of apatite, giving increased P activity in the soil solution. At soil pH between 6 and 7, the solubility of apatite has increased many times the value for a soil with pH of 8. But a lower pH, in turn, increases the solubility of Fe and Al compounds which, when pH becomes sufficiently low, will precipitate with the P released to the solution. Thus, as soil ages (see Figure 7.1), there is a decrease in apatite P and a corresponding increase in Fc and Al phosphate. The increase appears more often in the Fe-P fraction than the Al-P fraction. During this transitional phase, the organic P fraction has been accumulating with humus until it occupies an important portion of the total P in the soil, perhaps as much as 50% of the total P in surface soils.

7.2.2   Forms in Strongly and Intensely Weathered Soils

Ultimate weathering of soils leads to the situation shown along the right side of Figure 7.1. Soils have become almost totally depleted of Ca, Mg, and other cations and the soil is very acidic. Phosphorus is adsorbed or precipitated by Fe and Al, and soil organic P has become depleted because low levels of available or soluble P has limited biological growth. In many humid tropical areas, weathering has reached this stage and has produced low activity clay (LAC) soils. It is common to find more than 50% of the effective cation exchange capacity (ECEC) is Al-saturated and that the soil contains a variable charge arising from amorphous Fe and Al oxides and hydroxides. As shown in Table 7.1, total P has decreased drastically by this stage of weathering. The ultimate weathering of soil leaves it acid (see Table 7.1, pH 4.8), with the Ca-P fraction depleted, with organic P very low, and with the majority of P present as Fe-P. At this stage of weathering,

Table 7.1 Change in Phosphorus Forms with Weathering in
Venezuelan Soils

pH	Phosphorus form, meq/kg soil				
	Total P	Organic P	Ca-P	Al-P	Fe-P
6.9	692	235	70	33	43
5.9	298	79	88	20	33
5.0	144	85	3	14	19
4.8	59	11	0	2	17

Source: Sanchez, 1976, adapted from Westin and de Brito, 1969.

soil fertility is very poor with plant growth limited by low available P and also
by high levels of exchangeable and soluble Al.

7.3 PHOSPHORUS IN FERTILIZED AGRICULTURAL SOILS

Additions of P to soils began long before the manufacture of commercial
fertilizers as we know them today. The use of manures, and perhaps rock phosphate and organic sources of P, began as early as recorded time. Thus, the final
form of P that exists in soils, depends upon the state of weathering of the soil
and the type and the quantity of P that has been added.

Soils will vary from the calcareous, relatively unweathered forms to highly
weathered tropical soils, such as Oxisols. To understand what form P will take
when soils are fertilized, we need to know the possible forms of P in soils. They
are shown in Figure 7.2. Each of these forms of P may coexist in a particular
soil and each is converted from one form to another by components passing
through the solution phase. These forms coexist because the rates of dissolution
of one compound and of precipitation of another are many times kinetically very
slow. Thus, the rate of either formation or dissolution of a particular form of P
may be extremely important even though there is no indication of rates in Figure
7.2.

7.3.1 Soil Solution Phosphorus

Phosphorus in solution may exist as both orthophosphate or as higher polymers. Phosphorus in the soil solution, however, will be almost exclusively as
orthophosphate, PO_4^{3-}. If polyphosphates are added to the soil solution in the
form of fertilizers, they will hydrolyze to orthophosphate in a relatively short
time. This reaction will be faster if enzyme activated, but it also proceeds by
chemical hydrolysis. It is expected that 50% of added polyphosphate will be
hydrolyzed to orthophosphate in nine to sixteen days. Therefore, for practical
purposes, orthophosphate is the form that is most important.

The individual species of orthophosphate that will be in solution vary with
pH as shown in Figure 7.3. In most agricultural soils, the pH will be between
4.0 and 9.0; hence, the ion species present will be $H_2PO_4^-$ and HPO_4^{2-}. The ion
species present is somewhat relevant to plant uptake of P, since it has been shown

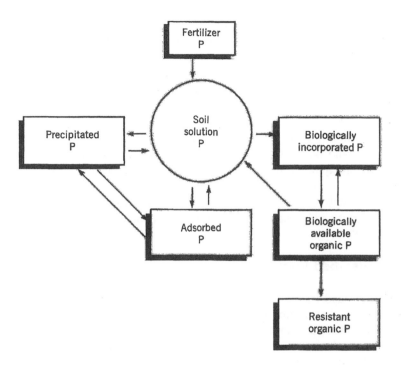

Figure 7.2 The soil phosphorus cycle.

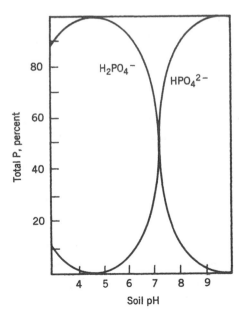

Figure 7.3 Mole percentage of each orthophosphate species as a function of pH.

that plants prefer the monovalent ion. But the rate of conversion of species in solution between HPO_4^{2-} and $H_2PO_4^-$ is so rapid that plants have little difficulty obtaining the necessary P for growth, even in soils with pH of 8 or higher, when levels of total P in solution are sufficiently high. Other ion pairs—made up of $CaHPO_4^0$, $MgHPO_4^0$, and $CaPO_4^-$ — may exist in soil solution if soil pH is above 7.0, but again the dissolution of these ion pairs will be rapid if the soluble HPO_4^{2-} is removed from solution.

The concentration of P found in soil solution may range from <0.01 to 7 or 8 mg/L, depending upon soil pH, recent additions of fertilizer P, and other soil factors. It would be unusual to find soil solution concentrations higher than 8 mg/L since precipitation will limit P to below this value in any soil pH range. In mineral soils, the maximum level of P in solution will be between pH 6.5 and 8. In Histosols with a very low mineral content, solution levels of P may increase with decreasing pH because the soils do not contain sufficient Fe or Al to precipitate P.

7.3.2 Adsorbed Phosphorus

The adsorbed fraction of soil P is often considered to be the *labile* soil P. Adsorbed P is that portion of soil P that is bonded to the surface of other soil compounds where a discrete mineral phase is not formed. For example, soluble P added to a soil solution may be bound to the surface of amorphous Al hydroxide without forming a discrete Al-P mineral. This would be an example of adsorbed P. Labile P is defined as that fraction that is isotopically exchanged with ^{32}P or is readily extracted by some chemical extractant or by plants. Thus, labile P may include some or all of the adsorbed P in a particular soil. On the other hand, in some soils the labile P may also include some precipitated P, or it may not include all of the adsorbed P. Labile P has been an important working concept for the soil scientist in relating soil P to plant available P. It is a measurable fraction, even though it may include P from several of the discrete fractions of P held in soils. Soil tests for P generally try to measure all or part of the labile P.

Two approaches to understanding adsorbed P are important in soil fertility. Models to describe the partitioning between liquid and solid phases of soils have been developed from laboratory measurements. These are very important because they describe how P will react in soils when broadcast and thoroughly mixed with them. This knowledge is necessary to predict how much P will be in solution for plant uptake and how much will be in solution that moves from the soil to runoff waters or will move down into the soil profile.

Soils vary greatly in their ability to adsorb P and in the relationship between solution P and adsorbed P. Extremes are illustrated in Figure 7.4. Generally, both very sandy soils and soils with a large amount of organic matter both hold only small quantities of P in the adsorbed form, and they do not form strong bonds with the adsorbed P. If we assume that 0.2 mg P/L is a desired quantity of P in soil solution for plant growth, 25 mg/kg (50 pounds/acre) of P would be adequate to raise the level from near zero to adequate for Grayling sand. The Brookston loam Ap horizon, which is more typical for agricultural soils, would require 61 mg/kg (122 pounds/acre) of P and the B horizon of the highly weathered Griffin soil would require 735 mg/kg (1,490 pounds/acre).

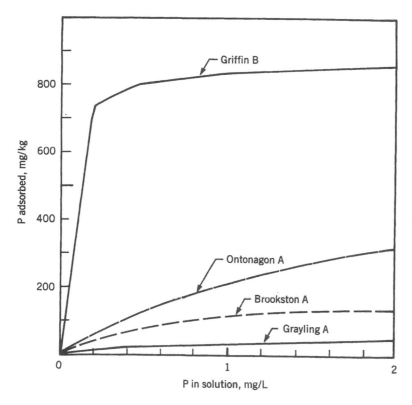

Figure 7.4 Partitioning of phosphorus between the solution and the solid phase of four soil horizons varying widely in texture and chemical properties. The texture of Griffin B and Ontonagan A is clay, of Brookston A loam, and of Grayling A sand.

During the past 100 years when modern agriculture and the fertilizer industry were developing, there was little concern about the consequences of high levels of adsorbed P in agricultural soils. Indeed, the greater problem during the early part of this period was how to obtain maximum uptake of P with limited application of P fertilizer. But this situation has changed drastically for heavily fertilized soils (see Table 7.2). With greater amounts of adsorbed P has come greater amounts of P in the water that runs off agricultural lands and percolates through soils. To this point in time, because most subsoils have a great capacity to adsorb P, there has been little increase in the P content of agricultural drainage waters and of groundwater. Exceptions may occur for sandy soils with a low capacity to adsorb P and have a short distance to groundwater. Also, tile-drained soils have a relatively little distance between the soil surface and the tile drain, although most soils requiring tile drainage have higher P adsorption capacities than sandy soils. In either case, long-time heavy application of P may result in movement of P to either surface or groundwater.

Table 7.2 Median Phosphorus Soil Test Values of Mineral Soils for Each of Six Geographical Regions in Michigan

Region	Median P soil test value, pp2m				
	1972	1976–1977	1979–1980	1985–1986	1990
Northern lower	42	44	89	82	85
West central	83	104	168	157	102
Thumb and eastern	41	55	79	99	76
South central	44	46	86	93	83
Southwest	72	98	143	132	130
Upper peninsula	36	32	77	63	44

Source: Mients and Warncke, 1983; and J. Dahl (personal communication).

A number of equations have been developed to describe quantitatively the distribution of P between the adsorbed and solution phases. The two most common are based on well-known equations formulated by Henry Freundlich and Irvin Langmuir for describing the volume of gas adsorbed as a function of pressure. According to the Freundlich equation:

$$Pad/m = k(Psol)^{1/n} \qquad (7.1)$$

where Pad/m = P adsorbed per m grams of soil, $Psol$ = P concentration (or activity) in the equilibrium solution, and k and n are constants defined experimentally for a given soil.

The use of this equation may be illustrated as follows. If k and n have been experimentally determined as 500 and 3, respectively, for a given soil, the quantity of P that must be adsorbed to give 0.2 ppm P in the soil solution is calculated as follows:

$$\log Pad = \log 500 + 1/3 \log 0.2 \qquad (7.2)$$

$$\log Pad = 2.70 + 1/3\,(-0.699) = 2.467 \qquad (7.3)$$

$$Pad = 293 \text{ mg/kg soil} \qquad (7.4)$$

The Langmuir equation is:

$$Pad/m = k(Psol)b/[1 + k(Psol)] \qquad (7.5)$$

where $P\,ad/m$ = P adsorbed per m grams of soil, $Psol$ = P concentration (or activity) in the equilibrium solution, b = the maximum P that can be adsorbed, k = a constant related to bonding energy.

A useful form of this equation is: $k = \theta/P(1 - \theta)$, where θ is the fraction of the adsorption maximum that is occupied by P. Again, once b and k are experimentally determined it is easy to calculate the quantity of P that must be adsorbed by a soil to raise the level in solution to a desired level.

For the four soils shown in Figure 7.4, it would require 1,470, 150, 122, and 50 pounds of P per acre for Griffin B, Ontonagon A, Brookston Ap, and Grayling A horizons, respectively, to raise the P level in solution from 0 to 0.2 mg P/L, a value which is necessary for normal plant growth. From another point of view, this is the quantity of P that each of the soils would adsorb from wastewater application before the soil solution level would exceed 0.2 mg P/L in solution.

The chemical form of adsorbed P would also be useful information, since this knowledge would help us to determine what minimum amount of P must be applied to increase availability. Components of soils which are important in adsorbing P are clays, Fe and Al oxides and hydroxides, and $CaCO_3$. Iron coatings that are associated with clay minerals may account for much of the clay's ability to adsorb P. Amorphous oxides of Fe and Al are much more effective at adsorbing P than crystalline compounds. As soils weather, they become able to adsorb large quantities of P because they have formed amorphous Fe and Al oxides and hydroxides. But as the weathering process continues, these amorphous compounds crystallize, exposing less surface area and fewer hydroxyls. They therefore adsorb less total P.

Calcium carbonate is a principal adsorption site for P in calcareous soils, but it has been shown that Fe compounds also function in adsorption in calcareous soils. Phosphate adsorbed onto $CaCO_3$ may later crystallize to precipitated P compounds.

7.4 PRECIPITATED PHOSPHORUS

Many insoluble compounds containing P form in soils. The principal compounds that are found or are precipitated when soluble P fertilizers are added to soils depend on soil pH and the other ions that are available for precipitation. We will discuss three groups of soils based upon their pH.

7.4.1 Calcareous Soils

Unless amounts of soluble P have been added to a calcareous soil, apatite P is expected to be the predominant precipitated form. Since it is a complex molecular form and crystalline in nature, apatite is so slow to dissolve that it does not furnish sufficient soluble P to maintain a soil in a fertile condition. Levels of P in soil solution depend on pH when apatite controls the solubility and are less than 0.01 mg P/L (10 ppb) for soil pH greater than 7. When the solution level of P is increased by addition of soluble P, for example by adding 0-45-0, dicalcium phosphate dihydrate, $CaHPO_4 \cdot 2H_2O$, forms within a few minutes. This compound is not a stable phase in soils and within three days anhydrous dicalcium phosphate, $CaHPO_4$, forms. It has been shown that octacalcium phosphate, $Ca_4H(PO_4)_3 \cdot 2.5H_2O$, forms in two to three months. Dicalcium phosphate and octacalcium phosphate are very significant in maintaining soil fertility of calcareous soils since both maintain much more soluble P than does apatite. Conversion of the phosphate to less soluble phosphates, such as tricalcium phosphate or apatite, requires many months or years. It has been shown that beta-tricalcium phosphate becomes the stable phase in calcareous soils that have not been fertilized with P for more than a year. The practical significance of this is that we can fertilize calcareous soils and expect to maintain suitable levels of P for plant growth for several months or even years even though much of the P precipitates.

7.4.2 Mildly Acid Soils

In the pH range of 5.5 to 7, adsorption may be more important than precipitation in controlling the level of P in soil solution. It is assumed that soluble P materials which precipitate in this pH range will precipitate as Fe or Al phosphates. The precipitation products that form will involve many complex species. Potassium and NH_4^+ taranakites, $H_6(K,NH_4)_3Al_5(PO_4)_8 \cdot 18H_2O$, form as well as mixed species including Ca, Mg, Al, and K. Simple compounds such as strengite, $FePO_4 \cdot 2H_2O$, and varisite, $AlPO_4 \cdot 2H_2O$, are likely to form only in very acid soils.

7.4.3 Strongly Acid Soils

Strongly acid soils are found in many areas of the world, including the southeastern U.S. and humid tropical areas of South America, Africa, and southeastern Asia. It would be a mistake to assume that all of these soils have similar properties. But many do possess sufficient soluble and exchangeable Fe and Al to precipitate P. Although strengite and varisite become good model compounds for precipitated P in the strongly acid soils, much of the P in these soils will be occluded in amorphous Fe and Al oxides and hydroxides. Fractionation schemes, which include an occluded or reductant-soluble fraction, will show the predominant fraction of P to be in this form. Identifying specifically precipitated compounds is extremely difficult, and so is distinguishing between the adsorbed and precipitated phases. Phosphorus will be strongly held in these soils, and the native soil fertility will be very limited due to low P solubility. These acid soils require careful management with respect to P. They can be quite productive if adequate P fertilizer, as well as other fertilizer nutrients, are utilized. Generally, band placement of P fertilizer is useful because plants will uptake P when economical P rates are used and P is band placed rather than incorporated. Levels of P may be built up in these soils after repeated fertilizer application.

7.5 BIOLOGICALLY INCORPORATED PHOSPHORUS

Plants absorb P directly from the soil solution. The total quantity removed by crops per year is small compared to the N and K removed, ranging from 6.0 to 26 kg/ha (5 to 23 pounds/acre) of P, with meadow hay absorbing small amounts and mangolds large amounts. In most agronomic plants, the P content will be between 0.1 and 0.4%. The concentration in plant tissue decreases considerably with age; for example, leaves from corn plants one foot high contained 0.48% P and the leaves from corn plants at tasseling contained 0.22% P. Excessively high values of P in plant tissue have been reported to be as much as 1%, but these were associated with P accumulation due to a deficiency of another nutrient, such as Zn.

Modern high-yielding varieties of plants have increased the level of P absorbed by plants. For example, three hybrids — Pioneer 3780, Michigan 5922, and Pioneer 3572 — yielding an average of 13.04 mg/ha (208 bushels/acre) of grain and 8.32 mg/ha (5.82 tons/acre) stover removed an average of 26.3 kg P/ha in grain and 4.7 kg P/ha in the stover. Alfalfa is a high accumulator of P and is expected to remove 59 kg P/ha for a 22 metric ton annual yield (10 tons/acre).

Since P is accumulated in the reproductive portion of the plant, which is usually harvested, the residue returned to the soil is often low in P. The ratio of P in the harvested portion to the residue is usually greater than 2:1. Although the fraction returned in the residue is important, it is normally less than 1% of the total organic soil P. Soil microorganisms will also compete for available P, but little data are available about the quantity of P that may be incorporated into living microorganisms. It is expected that microorganisms may function more in turnover of organic P than in retention.

7.6 SOIL ORGANIC PHOSPHORUS

As shown in Figure 7.2, organic soil P will include both biologically available organic P and resistant organic P. The exact chemical nature of organic soil P has been difficult to identify precisely. One form, inositol P, has been measured and is the largest identifiable fraction of organic P (see Table 7.3). The structure of one inositol compound is shown in Figure 7.5. There can be from one to six phosphate groups attached to the ring structure shown, which gives a number of slightly different compounds, all referred to as inositol P, all with similar properties.

Table 7.3 Inositol Phosphorus Content of Soils

Country	Inositol P, % of organic P	Organic P, % of total P
Denmark	46	61
New Zealand	5–26	30–77
Scotland	24–58	22–74
Canada	10–30	9–54
U.S.	10–25	3–52

Source: Halsted and McKercher, 1975.

The other known organic compounds, such as phospholipids, nucleic acids, phosphoproteins, and sugar phosphates, may be present but usually represent a small percentage of the organic P. The majority of the organic P present is a part of the stable humus material formed in soils. Like the N in humus, it is stable

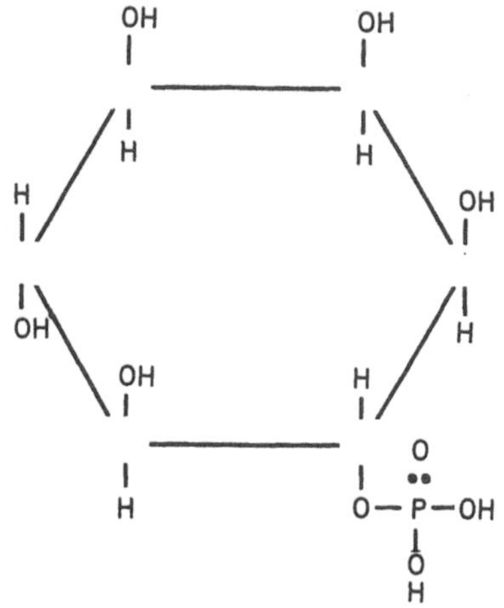

Figure 7.5 Structure of inositol monophosphate.

but difficult to identify as a precise organic compound; it is released slowly as the humus decomposes in the soil.

Organic P is important in supplying P for plant growth when fresh organic materials are added to soils. From 6 to 16% of labeled P added in the form of fungal mycelia was taken up by wheat plants in a five-week growing period. Uptake from an inorganic P source under similar growing conditions was above 20% of the P added. When manure is added as the source of organic P, the P is slowly released over a longer portion of the growing season, for the P in manure has been somewhat stabilized prior to application because microbes act on the less-stable organic P compounds in fresh manure.

The turnover of fresh organic P undoubtedly releases P to the soluble and labile pool. As such, it may be very important in furnishing available P for crops. However, the P released may be precipitated, absorbed by microorganisms or plants. The portion of this P that is absorbed by plants will depend upon the competition from the other processes that remove P from solution. Much of the organic P in the soil is in a form resistant to decomposition and is not readily available for biological uptake.

7.7 PHOSPHORUS UPTAKE BY PLANTS

Some studies with excised roots indicated that plants preferred the monovalent anion $H_2PO_4^-$ over the divalent HPO_4^{2-} by about 10 to 1. But since the conversion between the two species in solution is very rapid, this is probably of little importance for soils in the pH range 4 to 8.

Movement of P to plant roots is generally by diffusion rather than by mass flow. The contribution of mass flow to movement of P to roots can be easily calculated if assumptions are made about the transpiration ratio, the average concentration of P in plant tissue, and the average concentration of P in soil solution. For example, if we assume that 350 grams of water are used to produce 1 gram of dry plant tissue (transpiration ratio of 350 to 1), that the plant tissue concentration is 0.25% P, and that the soil solution contains on the average 0.1 mg P/L, then 350 ml × 0.1 mg P/L = 0.035 mg P, which will move with the water. One gram of tissue contains 2.5 mg P; therefore, 1.4% of the P would reach roots by mass flow. Since the level of P in solution is generally less than 0.2 mg/L, mass flow will not play an important part in P movement to roots in other than highly fertilized soils.

It has long been known that mycorrhizal fungi will increase the availability of nutrients to many plants. The mycelia of the fungi will become intimately associated with the plant roots and function in a symbiotic relationship. The mycelia become an extension of the roots. The increased availability of P to the host plant due to the mycorrhizal fungi is well known. Although better documented with tree species, numerous studies of agronomic crops have also shown greater plant growth, nutrient uptake, and yields due to mycorrhiza associated with plants growing in low-fertility soil. But evidence has shown that well-fertilized agronomic crops will out-yield unfertilized crops growing in low-fertility soil and have mycorrhizal fungi associated with their roots. Agronomic crops grown in less-developed countries where natural P levels are very low are expected to benefit more from P taken up by the mycelia of mycorrhizal fungi than do highly fertilized crops in developed countries.

7.8 BASIS OF PHOSPHORUS FERTILIZER RECOMMENDATIONS

Recommendations for P fertilizer are made on the basis of a soil-testing program in which P soil tests have been correlated with field fertility studies. Preliminary evaluations of a particular soil test for P or comparisons of different soil tests may involve greenhouse studies. During the past fifty or more years, many different extractants have been developed in attempts to measure labile P or P in the soil that would correlate with yield response to P fertilization. Three extracting solutions in widespread use will be discussed here. Others may be equally effective in local situations if properly correlated with field trials, and additional extractants may be developed in the future.

In 1945, Bray and Kurtz developed two extractants for P in soils, one for reserve P and one for available P. The available P test (Bray-Kurtz P1) utilized 0.025 N HCl + 0.03 N NH$_4$F as the extracting solution. Soil to solution ratios have varied from 1:7 to 1:50, and shaking time has varied from one to thirty minutes. Most soil testing laboratories in the north-central region of the U.S. are now using a 1:10 soil to solution ratio with a five-minute extraction period. A ratio of 1:50 with a one-minute extraction period has been shown to be much more effective on soils that are calcareous or have received applications of rock phosphate. But the procedure utilized in soil-testing laboratories has been very satisfactory in distinguishing between small, medium, and large amounts of available P.

Olsen developed the sodium bicarbonate extractable P test for use on calcareous soils. The P is extracted with 0.5 M NaHCO$_3$ buffered at pH 8.5 with a soil-to-solution ratio of 1:2 and a thirty-minute shaking or extracting time.

The more acid soils of the Piedmont area of the southeastern U.S. as well as acid, highly weathered soils of other regions of the world contain very little P that is extracted with either Bray-Kurtz P1 or Olsen's test. For these soils, Melich developed a stronger acid extractant utilizing 0.05 N HCl + 0.025 H$_2$SO$_4$ and later an improved extractant using 0.2 N CH$_3$COOH, 0.2 N NH$_4$Cl plus 0.015 N NH$_4$F plus 0.012 N HCl. However, in certain cases, the Melich 1 will extract less P than either the Bray-Kurtz P1 or the weaker Melich 3. This may occur because the latter extractants are able to displace P adsorbed by Al oxides and hydroxides. Melich 1 is more likely to dissolve apatite phosphates, which would be very low in highly weathered soils.

Analytical determination for P in solution is generally by the Murphy-Riley method utilizing the molybdenum blue color with ascorbic acid as the reducing agent. This method is widely adaptable to colorimetric methods including autoanalyzers and flow injection analyzers.

Most extractants utilized in soil tests do not remove the precise fraction of the nutrient in the soil that is absorbed by plants during a growing season. This does not make the soil test less useful or desirable. But the fraction that is extracted must correlate with that amount removed by plants during the growing season. Thus, in interpreting a soil test, the most important factor is that it correlate with uptake of the nutrient and with yield response to the nutrient. Generally, P soil tests are evaluated as low, medium, high, and very high. Soil tests that are low will require additions of P fertilizer that exceed the quantity removed by the crop being grown because P added in the fertilizer becomes fixed. Soils testing medium in P will require applications of slightly more than that removed by the crop for adequate yield and growth. When a soil test is high, no yield response is expected to P fertilizer. Here different approaches to fertilizer management may be followed. First, a starter fertilizer banded at planting time may be used to apply a maintenance amount of P. The intent is to apply the quantity of P that is removed by the growing crops so that the high soil test will be maintained. Alternately, no P fertilizer is applied and the soil is retested at frequent intervals to determine when P must again be applied.

When a soil test is very high in P, the probability of P being lost to surface waters through runoff and erosion is great. To reduce the risk of environmental degradation, fertilizer P should not be applied to soils testing very high in P. A possible exception to this is for potato production, a crop which sometimes gives yield responses to P fertilizer even though soils test very high. The soil should be retested frequently to determine when P levels are reduced to the point of requiring addition of P again.

7.9 RECOMMENDATIONS FOR PHOSPHORUS FERTILIZATION OF CROPS

Since fertilizer recommendations are specific to the area and crop, only general guidelines will be given here. They must be modified to fit soil type, the climate, and the yield goal of the farmer for a particular area. It is important for a farmer to have realistic yield goals when planning a fertilizer program. Fertilizing for 200 bushels of corn per acre is futile and expensive if the soil and climate will support only 120 bushels/acre.

From Table 7.4, it is apparent that soybeans and corn require considerably less P than does wheat. But the major difference comes with potatoes and other vegetable crops. Although it is easy to adjust and fertilize a crop that has a high requirement for P, the danger of excessive residual P in these soils is great and this leads to environmental concerns.

Table 7.4 Guidelines for Phosphorus Recommendations for Several Agronomic* Crops

Soil P test, mg/kg	Phosphorus applied, pounds/acre*			
	Corn	Wheat	Soybeans	Potatoes
15	30	39	22	76
30	13	22	0	70
50	0	0	0	60
80	0	0	0	48
175	0	0	0	0

* Yield goals assumed: corn, 140 bushels/acre; wheat, 70 bushels/acre; soybeans, 50 bushels/acre; and potatoes, 400 cwt/acre.

7.10 ENVIRONMENTAL CONCERNS OF SOILS HIGH IN PHOSPHORUS

Levels of P in surface waters greater than 10 ppb (10 µg P/L) have been associated with increased algae growth in streams and lakes. Continued high levels of P input to surface waters has led to eutrophication. Although agriculture may be a contributor to surface water degradation through inputs of P from heavily fertilized fields, this need not be the case in the future. As P accumulates in soils through additions of fertilizer P or organic materials such as manures, the surface soil will become very high in labile and total P and lead to movement downward in the soil profile or will increase the P content of water and soil particles transported to surface waters through runoff and erosion.

Responsible farm managers must be aware of how to control P losses to prevent damaging the environment. First, runoff water and soil erosion are the principal means of moving P from agricultural lands to surface waters. Conservation tillage and no-till systems that control erosion and water runoff are effective in reducing P loss and these management practices are essential when surface soils become high in available P. The small quantities of water lost through runoff when no-till is practiced may have an increased P concentration relative to conventional tillage, but the total quantity of P lost during the cropping season will be much less because the quantity of water that is lost is greatly reduced. The use of grass filters or riparian zones may also be beneficial in reducing the quantity of P lost to surface waters. Second, soil tests should be utilized even when soils contain high levels of P to define when P should not be added. Continued application of fertilizer P when soil tests are high is not a sound practice economically or environmentally.

Movement of P to lower horizons in a soil when the surface P levels are high does occur. Frequently there is ample depth of soil to remove the P from solution before it moves into tile drainage or groundwater. But the threat of environmental degradation by this route does exist if P application is continued for many years on soils testing high in available P.

REFERENCES

Bowman, R. A. and S. R. Olsen. 1979. A Reevaluation of Phosphorus-32 and Resin Methods in a Calcareous Soil. *Soil Sci. Soc. Am. J.* 43:121–124.

Bray, R. H. and L. T. Kurtz. 1945. Determination of Total, Organic and Available Forms of Phosphorus in Soils. *Soil Sci.* 59:39–45.

Cory, R. R. 1981. Adsorption versus Precipitation. *Adsorption of Inorganics at Solid-Liquid Interfaces*, pp. 161-182. M. A. Anderson and A. J. Rubin, Eds. Ann Arbor Science, Ann Arbor, MI.

Ellis, B. G. 1985. Phosphorus Cycle and Fate of Applied Phosphorus. In *Plant Nutrient Use and the Environment*. Kansas City, MO.

Ellis, B. G., C. J. Knauss and F. W. Smith. 1956. Nutrient Content of Corn as Related to Fertilizer Application and Soil Fertility. *Agronomy J.* 48:455–459.

Fixen, P. E. and A. E. Ludwick. 1982. Residual Available Phosphorus in Near-neutral and Alkaline Soils: II. Persistence and Quantitative Estimation. *Soil Sci. Soc. Am. J.* 46:335-338.

Gascho, G. J., T. P. Gaines and C. O. Plank. 1990. Comparison of Extractants for Testing Coastal Plain Soils. *Comm. in Soil Sci. and Plant Anal.* 21:13–16.

Griffin, R. A. and J. J. Jurinak. 1973. The Interaction of Phosphate with Calcite. *Soil Sci. Soc. Am. Proc.* 37:847–850.

Gross, D. W. and B. A. Stewart. 1979. Efficiency of Phosphorus Utilization by Alfalfa from Manure and Superphosphate. *Soil Sci. Soc. Am. J.* 43:523–528.

Halsted, R. L. and R. B. McKercher. 1975. Biochemistry and Cycling of Phosphorus. In *Soil Biochemistry*. E. A. Paul and A. D. McLaren, Eds. Marcel Dekker, New York.

Howeler, R. H., C. J. Asher and D. G. Edwards. 1982. Establishment of an Effective Endomycorrhizal Association on Cassava in Flowing Solution Culture and its Effects on Phosphorus Nutrition. *New Phytol.* 90(2):229–238.

Jensen, A. 1982. Influence of Four Vesicular-arbuscular Mycorrhizal Fungi on Nutrient Uptake and Growth in Barley. *New Phytol.* 90(1):45–50.

Juo, A. S. R. and B. G. Ellis. 1968. Particle Size Distribution of Aluminum, Iron and Calcium Phosphates in Soil Profiles. *Soil Sci.* 106:374–380.

Kapoor, K. K. and K. Haider. 1982. Mineralization and Plant Availability of Phosphorus from Biomass of Hyaline and Melanic Fungi. *Soil Sci. Soc. Am. J.* 46:953–957.

Kim, Y. K., E. L. Gurney and J. D. Hatfield. 1983. Fixation Kinetics in Potassium-aluminum Orthophosphate Systems. *Soil Sci. Soc. Am. J.* 47:448–454.

Lindsay, W. L. 1979. *Chemical Equilibria in Soils.* John Wiley & Sons, New York.

Lindsay, W. L. and H. F. Stephenson. 1959. Nature of Reactions of Monocalcium Phosphate Monohydrate in Soils: I. The Solution That Reacts with the Soil. *Soil Sci. Soc. Am. Proc.* 23:12–18.

Mehlich, A. 1984. Mehlich 3 Soil Test Extractant: Modification of Mehlich 2 Extractant. *Commun. in Soil Sci. and Plant Anal.* 15:1409–1416.

Meints, V. and D. D. Warncke. 1983. Changes in Soil Test Levels of Mineral Soils in Michigan, 1962-1968. *Bull. for Crops and Soils Ext. In-service Training.* Michigan State University, East Lansing, MI.

Murphy, J. and J. P. Riley. 1962. A Modified Single Solution Method for the Determination of Phosphate in Natural Waters. *Anal. Chim. Acta* 27:31–36.

Olsen, S. R. and F. S. Watanabe. 1970. Diffusive Supply of Phosphorus in Relation to Soil Textural Variations. *Soil Sci.* 110:318–327.

Olsen, S. R. and F. S. Watanabe. 1957. A Method to Determine a Phosphorus Adsorption Maximum of Soils as Measured by the Langmuir Isotherm. *Soil Sci. Soc. Am. Proc.* 21:144–149.

Olsen, S. R., C. V. Cole, F. S. Watanabe and L. A. Dean. 1954. Estimation of Available Phosphorus in Soils by Extraction with Sodium Bicarbonate. *U.S.D.A. Cir. 939.*

Pacovsky, R. S., G. J. Bethlenfalvay and E. A. Paul. 1986. Comparisons Between Phosphorus Fertilized and Mycorrhizal Plants. *Crop Sci.* 26:151–156.

Ryan, J., D. Curtin and M. A. Cheema. 1985. Significance of Iron Oxides and Calcium Carbonate Particle Size in Phosphate Sorption by Calcareous Soils. *Soil Sci. Soc. Am. J.* 49:74-76.

Ryan. J. and M. A. Zghard. 1980. Phosphorus Transformations with Age in a Calcareous Soil Chronosequence. *Soil Sci. Soc. Am. J.* 44:168–169.

Sample, E. C., R. J. Soper and G. J. Racz. 1980. Reactions of Phosphate Fertilizers in Soils. In *The Role of Phosphorus in Agriculture.* F. E. Khasawheh, E. C. Sample, and E. J. Kamprath, Eds. Am. Soc. Agronomy, Madison, WI.

Sanchez, P. A. 1976. *Properties and Management of Soils in the Tropics.* John Wiley & Sons, New York.

Sharpley, A. and M. Meyer. 1994. Minimizing Agricultural Nonpoint-Source Impacts: A Symposium Overview. *J. Environ. Qual.* 23:1–3.

Sposito, G. 1980. Derivation of the Freundlich Equation for Ion Exchange Reactions in Soils. *Soil Sci. Soc. Am. J.* 44:652–654.

Sposito, G. 1979. Derivation of the Langmuir Equation for Ion Exchange Reactions in Soils. *Soil Sci. Soc. Am. J.* 48:336–340.

Stowasser, W. F. 1983. *Phosphate Rock. Mineral Commodity Profiles.* U. S. Dept. of Interior, Bureau Mines, Washington, D.C.

Tiessen, H. J., W. B. Stewart and C. V. Cole. 1984. Pathways of Phosphorus Transformations in Soils of Differing Pedogenesis. *Soil Sci. Soc. Am. J.* 48:853–858.

Westin, F. C. and J. C. de Brito. 1969. Phosphorus Fractions of Some Venezuelan Soils as Related to Their Stage of Weathering. *Soil Sci.* 107:194–202.

Wilson, M. A. and B. G. Ellis. 1984. Influence of Calcium Solution Activity and Surface Area on the Solubility of Selected Rock Phosphates. *Soil Sci.* 138:354–359.

Withee, L. V. and R. Ellis. 1965. Change of Phosphate Potentials of Calcareous Soils on Adding Phosphate. *Soil Sci. Soc. Am. Proc.* 29:511–514.

CHAPTER 8

Potassium

Of the elements essential for plants, potassium and calcium are the two most abundant in the earth's crust. Potassium accounts for 2.6% of the earth's crust, and calcium 3.6%. It has been estimated that the average plow layer in the U.S. contains 0.83% K, which equals 16,600 pp2m. Most of the soils in the western half of the U.S. are minimally or moderately weathered and contain 1.7 to 2.5% K. By contrast, many soils of the southeastern coastal plains contain only 0.3% K. Therefore, many soils have high amounts of total and plant-available K, and responses to K fertilizers are nonexistent or low. On the other hand, the contents of many soils are low in both total and plant-available K, and yield responses to K fertilizers are high. Over 80% of the K fertilizer applied in the U.S. is used in the Lake states, the Corn Belt, Appalachia, and the Southeast, areas with udic soil moisture regimes.

Potassium is not complexed or bound up into organic matter to any degree, compared to N; essentially all of it is associated with the mineral fraction. For this reason, soil K and soil N can be considered opposites. Soil N exists mainly as organic N, and a minor amount exists as mineral N. Soil K is mostly mineral, and the daily K needs of plants are little affected by organic associated K, except for exchangeable K adsorbed on organic matter. As a result, mineralization of organic N is the major source of plant-available N and the weathering of K minerals is the major process for converting unavailable K into forms that plants can use.

8.1 THE SOIL POTASSIUM CYCLE

Soil parent material contains K that is mainly in feldspars and micas. These minerals weather, and the K ions released are exchangeable and exist as adsorbed and solution K. From the soil solution, the K^+ may undergo various changes in the soil including fixation and uptake by plants or micoorganisms, after which they may return again to the soil solution. In the humid regions, K is lost from the soil by leaching as shown in Figure 8.1. A potassium ion could, hypothetically,

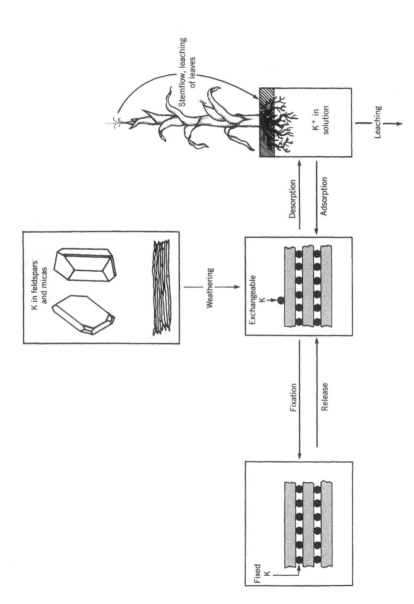

Figure 8.1 Major segments of the soil potassium cycle.

be released from primary minerals by weathering and occur as an exchangeable cation. It could become fixed, released from fixation, and become an exchangeable cation again; desorb into the solution; be absorbed by roots, leached from leaves, and returned to the soil solution again; and then be carried to the groundwater by leaching or removed from solution by being adsorbed as an exchangeable ion or fixed in interlayer positions of clay minerals.

The exchangeable K, XK, maintains an equilibrium with the K in solution, and together the exchangeable and solution K make up the available K. The solution K is equal to 1 to 3% of the XK in most soils. The time required for equilibration of solution and adsorbed K is usually less than an hour, and on the order of days or weeks for equilibration between exchangeable and K fixed in interlayer positions. Depletion of solution K by plant uptake or leaching causes adsorbed K to quickly move into solution to reestablish the equilibrium. Since the adsorbed K tends toward an equilibrium with the fixed K, some fixed K will slowly become exchangeable. The reverse is true when the solution K is increased by K fertilizer (see Figure 8.1).

Because the mineral component in organic soils or Histosols is limited, most of the K exists as XK, and the soil has little or no capacity to fix K. Mineral materials tend to occur in Histosols as contaminants and the amount of mineral material affects the K relationships in Histosols.

With time, theoretically, all of the primary mineral K is released and eventually removed from the soils in humid regions by leaching. The most-weathered soils have few weatherable minerals in the coarse silt and fine sand and are, in general, infertile soils with little capacity to supply K (and Ca, Mg, etc.) for plant growth. In the larger K cycle, the K ions leached from the soil appear in the oceans where they encounter unweathered and partially weathered minerals, including 2:1 clays. Potassium ions re-enter the structures of these deeply buried micaceous minerals. These K minerals have been called sedimentary micas to differentiate them from primary mineral mica, biotite, and muscovite. When sediments containing these K minerals are uplifted for another cycle of weathering and soil formation, another K weathering cycle is initiated.

8.2 MINERAL WEATHERING RELEASE OF POTASSIUM

The amount of available K in unfertilized soils today depends on the amount and kinds of K minerals that were in the parent material and the kinds and degree of changes during soil genesis. Thus, K release in soils derived from igneous and metamorphic rocks will be restricted to the weathering of mica and feldspar, whereas soils containing sedimentary materials may have considerable K released from sedimentary mica, as well as feldspars and micas. Many fine-textured glacial sediments around the Great Lakes are rich in sedimentary mica derived from shales and limestone, whereas most of the glacial sediments on the Canadian Shield in northeastern Canada contain K minerals derived from granite and other granitic basement rocks.

8.2.1 Potassium Minerals

The major primary K minerals in soils, feldspars and micas, have the following chemical formulas:

Feldspar
  Microcline $KAlSi_3O_8$
  Orthoclase $KAlSi_3O_8$
Mica
  Muscovite $K(AlSi_3)Al_2O_{10}(OH)_2$
  Biotite $K(AlSi_3)(Mg,Fe^{2+})_3O_{10}(OH)_2$

The K content decreases from about 13% to 14 for feldspars to about 10% for micas. Feldspars and micas, along with quartz, are the major minerals in granitic rocks.

Feldspar particles in soils tend to be larger in size than mica particles. In some well-drained Mollisols developed from till in Saskatchewan, Canada, over 50% of the mica was in the clay with only a minor amount in the sand, but over 50% of the feldspar was in the sand and a minor amount in the clay (Figure 8.2). Finer-sized particles have greater specific surface and weather more rapidly than coarser particles. In most soils the micaceous minerals are more important than feldspars in replenishing the plant available K.

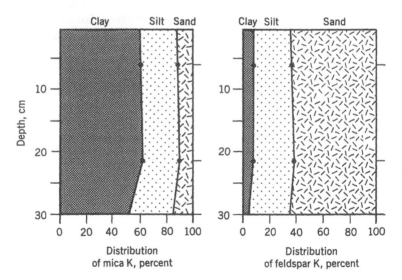

Figure 8.2 Distribution of mica and feldspar potassium among the sand, silt, and clay in a well-drained Mollisol (Orthic Black) in Saskatchewan, Canada. (Adapted from Somasiri, Lee, and Huang, 1971.)

Many sandy soils whose clay content is low have low fertility because most of the sand and silt particles are quartz. The most infertile soil in Wisconsin is Boone sand, which developed from quartzitic sandstone. If a soil has developed from arkose (feldspathic) sandstone, however, the soil could be sandy and provide much more available K than a soil like the Boone sand. The weathering of feldspar in the sand fraction contributes to the release of large amounts of K in some sandy Ultisols of the coastal plains in Delaware. As a general rule, fine-textured soils have greater K supplying power than coarse-textured soils. This greater supply of K is related, in part, to the fact that mica is nearly a universal constituent of the clay in soils.

8.2.2 Feldspar Weathering and Potassium Release

Quartz is composed of shared $Si-O_4^{4+}$ tetrahedra and the chemical formula SiO_2 of quartz could be written Si_4O_8. Compare this to the feldspar formula, $KAlSi_3O_8$. In a sense, one of every four of the tetrahedra has Al^{3+} substituted for Si^{4+} and the charge is balanced out by the inclusion of a K^+ in the feldspar structure. Feldspars are quite resistant to chemical breakdown and weather slowly; the microcline weathers more slowly than orthoclase. During weathering, there is a complete breakdown of the crystal structure and release of the component parts. The K^+ is released into the soil solution; however, some of the Al and Si form a gel that coats the weathering surface and forms a protective barrier that inhibits further dissolution of the feldspar. Although feldspar is an important source of K^+, its slow dissolution rate results in poor correlations between the amount of feldspar in a soil and the amount of K absorbed by plants.

8.2.3 Mica Weathering and Potassium Release

The structural and unavailable K in micas exists in the interlayer space (see Figure 2.6). Internal crystal-bonding differences cause trioctahedral biotite to weather more rapidly than dioctahedral muscovite. This allows K to leave the structure more easily in biotite than in muscovite.

The interlayer K is lost first along particle edges. Loss of interlayer K causes unbalanced negative charge in adjacent tetrahedral sheets of 2:1 layers, and their mutual repulsion causes expansion of the interlayer space in the weathered particle edges. It has been theorized that an interlayer wedge-shaped zone is created at the juncture of the unweathered mica core. The cation adsorption sites closest to the unweathered mica core have the smallest interlayer distance and have the greatest affinity for K. Small cations, such as H_3O^+ and NH_4^+, are the most effective in exchanging for structural K^+.

There is decreasing affinity for K in the wedge zone with increasing distance toward the particle edge, as the interlayer distance increases. Thus, a K availability continuum is created, consisting of unweathered core mica K and XK adsorbed to sites with different energies of adsorption. The unavailable K occurs in the i (interior) positions and the exchangeable K at the p (planar) and e (edge) positions as shown in Figure 8.3. The e position, near the site of a vacated structural K, has a special affinity for K^+ and holds XK more strongly than on the planar surfaces (p site). Since the various forms of K tend toward equilibrium with one another, they operate as buffers to moderate the rate at which solution K is depleted when K is absorbed by roots or leached.

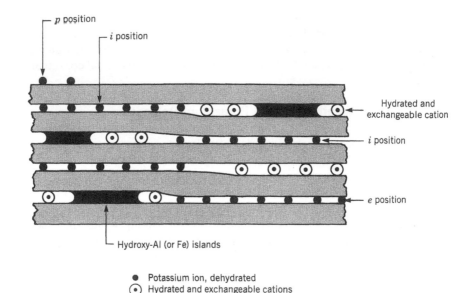

Figure 8.3 The exchangeable potassium is adsorbed at the *p* or planar position and at the edge or *e* position. Nonexchangeable potassium is fixed in the interlayer or *i* position. Strongly hydrated cations, like calcium, enter the interlayer space and remain exchangeable. Hydroxy-Al islands hold layers apart and may trap ions in the interlayer space. (After Rich, 1968. Used by permission of the Soil Science Society of America.)

Also note in Figure 8.4 that the formation of hydroxy-Al islands in the interlayer space blocks the loss of K from and reentry of K into the interlayer space.

During the normal course of soil development in humid regions, there is a gradual release of K from K bearing minerals and a net loss of K from the soil by leaching. Thus, many soil particles consist of mica cores with weathered zones of vermiculite and others consist mainly of mica or mainly vermiculite. Models illustrating the weathering sequence are shown in Figures 2.6a–c.

8.2.4 Weathering Rates and Relative Availability

The four primary K minerals were finely ground and used in an experiment to measure their ability to supply K for oats. The removal or uptake of K by the oats paralleled the weathering rate sequence, with biotite being the most easily weathered and supplying the most K; microcline weathered the slowest and supplying the least K for uptake. Removal of K by carbonated water extraction paralleled the plant removals (Table 8.1).

The relative weathering rates and release of K from microcline, muscovite, and biotite have been reported to be 1.0, 1.8, and 190, respectively. Many experimenters have found that biotite is a good source of K for growing plants. There is evidence that roots play a role in the removal of structural K from biotite and

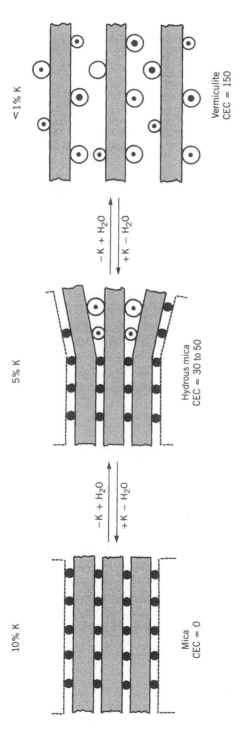

Figure 8.4 Schematic weathering of mica to hydrous mica and to vermiculite, and the reverse sequence caused by potassium fixation. (Modified from McLean, 1978).

Table 8.1 Availability of Potassium in Freshly Ground Minerals

Source of K	K removed by oats (gains over control), grams/pot	K removed by five extractions,* ppm
K_2SO_4	0.253	—
Biotite	0.203	217
Muscovite	0.177	141
Orthoclase	0.062	78
Microcline	0.011	51

* 30 grams in 200 ml of carbonated water shaken for 96 hours.

Source: Data of Plummer, 1918, and modified from Rich, 1968.

that plants differ in their capacity to extract or use biotite K. The rapid weathering rate of biotite accounts for its low content in most soils, whereas the very slow weathering rate of muscovite accounts for its presence in the clay fraction of most soils.

8.2.5 Weathering of Sedimentary Mica

Illite was proposed by Grim in 1937 as a name for mica occurring in argilla-ceous sediments in Illinois. The use of the term has been confusing and it is used here to refer to sedimentary mica that is heterogeneous and containing both nonexpanded and expanded minerals. Illite as a mixture of materials is a rock term as contrasted to a mineral term. Compared to mica, the K content is lower and the water content is higher. For general purposes, the weathering of illite and the release of K and its fixation parallel those for unweathered mica, weathered mica, and vermiculite.

8.3 POTASSIUM FIXATION

After treatment with a solution containing K, and leaching with neutral normal ammonium acetate, many soils retain some of the added K in nonexchangeable form. The retained K is fixed, mostly by immobilization in wedge zones and expanded interlayer spaces of minerals where K^+ originally existed before being weathered out. In field soils, the application of K fertilizer produces local pockets of high-K^+ concentration and promotes the re-entry of K into wedge zones and interlayer spaces. Re-entry or fixation of K into the structure of vermiculite and weathered mica converts them back to mica. The changes in K content, CEC, and structural relationships in the weathering and fixation in the mica-weathered mica-vermiculite sequence is illustrated in Figure 8.4.

8.3.1 Wedge-Zone Fixation

The strong negative charge near the apex of wedge zones, originating in the tetrahedral sheet and located where structural K^+ weathered out, preferentially adsorbs K^+ with an accompanying collapse of the expanded layer edges. The low energy of hydration and the ease with which the hydration water is lost facilitates movement of K^+ into the wedge-shaped space. This converts available K^+ into unavailable or fixed K. Fixation occurring near the apex of wedge zones, where bonds are satisfied over the shortest distance, fixes K by specific adsorption. Wedge-zone fixation in field soils is little affected by normal changes in soil moisture and is highly dependent on K^+ diffusion gradients. Fixation is rapid and subsequent release is slow, creating a mechanism for the conservation of soluble K in fertilized soils.

8.3.2 Expanded Interlayer Space Fixation

Potassium fixation occurs in the interlayer space of 2:1 expanding layer silicate minerals. The electrostatic forces between adjacent layers tend to cause collapse of the interlayer space and trap or fix cations while hydration of the interlayer space and hydration of interlayer exchangeable cations act to prevent interlayer space collapse and fixation. Charge located in the tetrahedral sheet is more effective in causing interlayer collapse and cation fixation than charge originating farther from the interlayer plane in the octahedral sheet. Drying produces dehydration and facilitates collapse of the interlayer space and fixation of K. In most field soils, the moisture content and other conditions are such that vermiculite does not fix a significant amount of K. Air drying of the immediate soil surface, however, is sufficient to fix K in field soils containing vermiculite. Oven drying can fix K in montmorillonite, which has lower charge density than vermiculite and dioctahedral charge.

Fixation of K in the laboratory by 44 clay samples removed from soils with smectitic, vermiculitic, or mixed mineralogy occurred under both wet and dry conditions and was strongly correlated with total CEC determined at pH 7 and the amount of tetrahedral CEC. Both tetrahedral and octahedral CEC were associated with fixation; however, the weighted effect of the tetrahedral charge CEC was 64% and the weighted effect of the octahedral charge CEC was 36%.

Drying soil samples before testing for exchangeable K has important implications for making K fertilizer recommendations. Air drying can produce K fixation in the interlayer spaces of vermiculite and oven drying fixes K in both vermiculite and montmorillonite. Some studies have shown release of K due to oven drying. One theory holds that drying vermiculite causes collapse of the interlayer space, while causing a gross expansion of vermiculite particles. The gross physical change seems to be due to curling of mineral layers resulting in greater opportunity for K release. The effect of the equilibrium status of the soil sample at the time that it is collected may also be important in determining if K is fixed or released upon drying. If the soil has been depleted of XK (at the end of a growing season), then it is likely to release K upon drying. But if the soil has been recently fertilized with K (at the beginning of a growing season), fixation of K upon drying is likely to occur. The effects of wetting and drying on K fixation and release need further elaboration.

Development of hydroxy-Al interlayers in 2:1 expanding clays inhibits interlayer collapse and greatly reduces K fixing capacity. These hydroxy-Al clays, however, may contain K trapped between islands of a hydroxy-Al matrix (see Figure 8.4). The weathering out of the hydroxy-Al interlayer releases entrapped K. Kaolinite and oxidic clays do not fix K.

8.3.3 Significance of Potassium Fixation

The importance of K fixation depends on the situation, especially the soil mineralogy. In some agricultural soils, K fixation conserves K because it reduces K loss by leaching. On the other hand, some agricultural soils have such large K-fixing capacities that large amounts of K fertilizer have little influence on yields. In Michigan, for example, crop yields were little increased, if at all, by a high level of K fertilization in a field that occupied a downslope position in the landscape. It was found that hydroxy-Al interlayered vermiculite clay had formed on the upper slopes in an acid environment and was moved downslope by erosion and deposited where soils were much more alkaline. The greater alkalinity removed much of the interlayer hydroxy-Al and increased the clay's K-fixing capacity. Much of the fertilizer K was satisfying the soil's K-fixing capacity and was not available for crop growth. Banding of K fertilizer in soils with high fixing capacity has been found to increase the effectiveness of K fertilizer. The pH effect, where increasing pH increases K fixation, appears to be due to the removal of interlayer hydroxy-Al and increased K fixation in the interlayer spaces.

High K-fixation rates by vermiculitic clay in some soils of the San Joaquin Valley of California greatly reduced the effectiveness of K fertilization of cotton. The fixation is correlated with reduced CEC of the vermiculitic clay.

The importance of the various forms of K, including fixed K, as a K source for plants is illustrated by the results of a greenhouse experiment in which nine consecutive crops of alfalfa were grown on thirteen different soils. The soils contained an average of 1.62% total K, equal to 36,210 pp2m. The amounts of exchangeable, fixed, and residual K (in micas and feldspars) were 1, 3, and 96% of the total, respectively (Table 8.2). The alfalfa crops removed 399 kg/ha (356 pounds/acre) with 44% from a depletion of XK, 39% from fixed K, and 17% from the residual K. The uptake of one unit of K was correlated with 2 units of XK, 7 units of fixed K, and 517 units of residual K. During this time period, K uptake was significantly correlated with exchangeable and fixed K, but not with residual K.

In extremely weathered tropical soils, there is generally little, if any, feldspar or mica K, and most of the K is exchangeable. Their kaolinitic and oxidic mineralogy results in little, if any, K fixation. Extreme K-deficiency symptoms are likely to develop when there is a large K demand for crop production (Figure 8.5). On the other hand, many Oxisols in Africa have developed from granitic basement rocks rich in mica and still contain enough mica (muscovite) and hydroxy-Al trapped K to provide sufficient K for low-intensity farming. Intensive crop production, however, requires high K fertilization.

Table 8.2 Amounts of Various Forms of Potassium in Thirteen Iowa Soils and Removal by Thirteen Consecutive Crops of Alfalfa

Form of potassium	Soil potassium			Potassium uptake		
	kg*/ha	pounds* per acre	percent	kg/ha	pounds per acre	percent
Exchangeable	358	326	1	176	157	44
Fixed	1,084	967	3	156	139	39
Residual	34,768	31,015	96	67	60	17
Total	36,210	32,302	100	399	356	100

* Kilograms or pounds based on 2,000,000 (kg or lbs) of soil.

Source: Data from Pratt, 1951.

Figure 8.5 Extreme deficiency of potassium on maize (corn) producing lodging of the maize grown on a Brazilian Oxisol.

8.3.4 Summary Statement

As soils evolve, their K-fixing capacity increases due to the weathering of mica and formation of weathered mica and vermiculite. The formation of hydroxy-Al interlayers in 2:1 layer expanded minerals inhibits K fixation. The eventual disintegration of 2:1 minerals with hydroxy-Al interlayers releases any K entrapped in interlayer spaces. The disintegration of 2:1 aluminosilicate layer clays and the formation of kaolinite and oxidic clays over time is associated with decreasing K-fixing capacity. Thus, K release and fixation tend to be moderate or high in minimally and moderately weathered soils with high-activity clay and K release and fixation tend to be very low in extremely weathered soils with low activity clay.

8.4 LEACHING LOSS

The loss of K by leaching is related to the concentration of K in the soil solution and the amount of water that leaches through the soil. The generally low concentration of solution K minimizes leaching losses. Assuming a K concentration of 4 mg/kg of water and the leaching of 25 cm of water, the loss is equal to 10 kg/ha or 9 pounds/acre. This loss is typical of soils having high-activity clays and is small relative to the uptake of high yielding crops. Leaching losses and extent of K movement tend to increase as soils become more sandy.

Significant leaching losses are more likely on soils with low activity clay due to low CEC and limited K fixation. As much as 25% of the K was lost by leaching during the first year when 500 kg/ha of K was applied to Oxisols on the savannah in Brazil. Normally, K application rates and leaching losses are much less. However, losses of K by leaching appear to be more serious on soils with low-activity clays than soils with high-activity clays, and K from fertilizer application moves more deeply. In arid areas where soils are only minimally or moderately weathered and little leaching occurs, K fertilizers are usually unnecessary.

8.5 PLANT AND SOIL POTASSIUM RELATIONSHIPS

Of the nutrients removed from soils, plant uptake of K is often second to N. It is not uncommon for annual crops to remove over 224 kg/ha or 200 pounds/acre of K (Figure 8.6).

Figure 8.6 The alfalfa for this record nonirrigated yield contained 3% potassium, an uptake of potassium equal to 600 pounds per acre or 673 kilograms per hectare. (Photograph courtesy Dr. M. B. Tesar.)

8.5.1 Uptake and Role in Plants

Potassium is absorbed from solution as an ion and plant uptake is closely related to solution concentration. Uptake is active and against a concentration gradient unless the solution concentration K is high. Potassium is associated with many enzymes involved in photosynthesis, organic compound synthesis, and translocation of organic compounds. The relative uptake of K early in the growing season is generally greater than that for nitrogen and phosphorus. Potassium is not complexed into organic compounds in plants and may be lost from plants late in the season due to leaching of foliage and leakage from roots. Thus, the amount of K in plants may be greater some time during the growing season than at harvest (see Figure 1.9).

In cereals, the K is concentrated in the vegetative parts so that the harvest of corn for silage removes large quantities of K as compared to just the removal of corn grain. Much of the K taken up to produce cassava is left behind after tubers have been harvested, allowing great potential to recycle most of the K taken up to be used by succeeding crops.

Potassium is mobile in plants. Deficiency symptoms appear on the oldest leaves as a chlorotic marginal firing. In advanced stages, the entire leaf turns yellow and is necrotic. The deficiency symptoms for many crops is a yellowing and eventually death of cells near leaf margins. Extreme K deficiency produces weak stalks and lodging of grain crops (see Figure 8.5).

Potassium deficiency symptoms on cotton were some of the first plant deficiency symptoms recognized. Older leaves develop a yellowish-white mottling. Leaves become yellow-green and yellow specks appear between leaf veins. The leaf tip and margins curl downward and, finally, the entire leaf becomes rust colored, giving rise to cotton rust, a term for K deficiency. Early leaf fall is common and boll development stops. Potassium deficiency symptoms are common in the southeastern U.S. because cotton is inefficient in removing K from soils, compared to many other plant species, and much of the cotton is planted on Ultisols with very low K supplying power.

8.5.2 Potassium Uptake from Soils

Various studies have shown that the amount of exchangeable K in many soils is about equal to the annual uptake of productive crops, and that the amount in solution at any given instant is equal to 1 to 3% of the exchangeable K. Assuming that the solution K is equal to 2% of the exchangeable K, the soil solution in effect will have to be depleted and replaced with K about 50 times during the growing season. Assuming that crop needs and exchangeable K in the plow layer are both 200 kg/ha (or about 200 pounds/acre) and that the solution K is equal to 2% of the exchangeable, 4 kg of K will be in solution per hectare. Daily maximum uptake rates of 8 to 10 kg/ha have been reported for corn. Thus, the solution would contain only a 12-hour supply during the period of maximum uptake. Obviously, mass flow will not move enough K^+ to the roots to supply plant needs.

Of the K taken up by corn, it has been estimated that 2% comes from root interception, 20% from mass flow, and 78% from diffusion. Most of the diffusion occurs within 4 mm of the root surface. The average K in the saturated extracts of 142 soils of the north-central region of the U.S. was found to be 4 mg/kg with a range of 1 to 80. For the soil solution to transport enough K to roots by mass flow, the concentration would need to be over 40 mg/kg.

The uptake of K is strongly correlated with the solution concentration of K ions which depends on the equilibrium relation between the solution K and exchangeable K, and between the exchangeable K and fixed K. Early in the season, plants deplete the solution K and solution K is rapidly replaced from the exchangeable K. As solution K is depleted with time, replacement of solution K slows down, for K ions exchange from sites having greater energy of adsorption. Less K comes from p sites and increasingly more K comes from specific adsorption sites on mica edges or in expanded sections of mica wedges. With time, the solution concentration of K and amount of exchangeable K decrease. The lowered XK encourages the movement of fixed K to exchange positions. These events create a pattern of a reduction of XK early in the season followed by a reduction in fixed K later in the season (Figure 8.7).

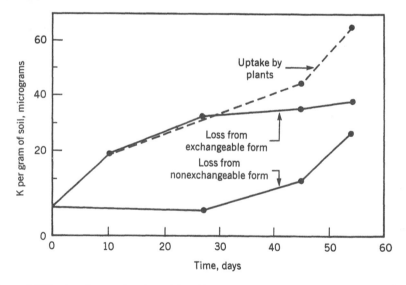

Figure 8.7 Uptake of potassium from 1 kg of loam soil by six corn plants in the greenhouse and loss of potassium from exchangeable and nonexchangeable forms in the soil. (From *Soil-Plant Relationships*, 2nd ed., C. A. Black, © 1968. Reprinted by permission of John Wiley & Sons.)

8.5.3 Luxury Consumption

Luxury consumption of K is seldom a problem. High application of K fertilizer, however, can result in excessive or luxury consumption of K. The results of an experiment, as shown in Table 8.3, illustrate how the uptake of K rises to luxury levels.

Table 8.3 Effects of Removal and Replacement of Exchangeable and Fixed Potassium on Yields and Potassium Uptake

Soil treatment	Dry matter yield			K uptake		
	kg*/ha	pounds* per acre	percent	kg/ha	pounds per acre	percent
Untreated	9,585	8,550	100	43	38	100
Exchangeable K removed	5,997	5,350	62	18	16	41
Exchangeable K and fixed K removed	1,973	1,760	21	10	9	26
Treated plus K fertilizer	14,259	12,720	149	413	368	968

* Kilograms or pounds based on 2,000,000 (kg or lbs) of soil.
Source: Adapted from Truog and Attoe, 1945.

A large quantity of field soil was collected and brought to the greenhouse and divided into three samples. One was untreated; a second was leached with a salt solution, which removed the XK; and the third sample was leached for several hours with HCl, which removed both the XK and the fixed K. These treatments were the first three listed in Table 8.3. Part of the soil from which the XK had been removed, by the third treatment, was then given a large application of K fertilizer, which restored the fixed K and the XK. The amount of K applied was equal to 800 pounds/acre; this procedure made up the fourth "treated plus K fertilizer" treatment. Crops were then grown on these parcels of soil. Removal of the XK reduced yields to 62% of the untreated soil yields. Removal of the XK and fixed K reduced yields to only 21% of untreated soil yields. Decreases in the uptake of K paralleled the removal of exchangeable and fixed K.

The addition of K fertilizer to soil that had both the exchangeable and fixed K removed produced a yield that was 149% of the untreated soil. By contrast, however, the K fertilizer resulted in such high levels of solution K that uptake was 968% of the untreated soil. Such an uptake constitutes luxury consumption of K. The yield increase was 149%; however, the increase in K uptake was 968%.

8.6 POTASSIUM BUFFER CAPACITY

Removal of K from the soil solution by plant uptake during the growing season is usually associated with a decrease in XK that is followed by a decrease in nonexchangeable K as shown in Figure 8.8. Changes in XK are associated with changes in the concentration of K in solution, that is, with K availability. The ability of the soil to maintain the solution concentration of K during the growing season with depletion of XK is the potassium buffer capacity, KBC. The

KBC can be ascertained from graphs in which are plotted the XK on the ordinate and solution K concentration on the abscissa. The slope of the line, or the ratio of the change in XK to the change in solution K concentration, is the KBC. A steep line reflects a high KBC where the soil is capable of maintaining the solution K, while experiencing a large decrease in exchangeable K. Well-buffered soils tend to have high CEC and contain large amounts of XK and mica wedge K.

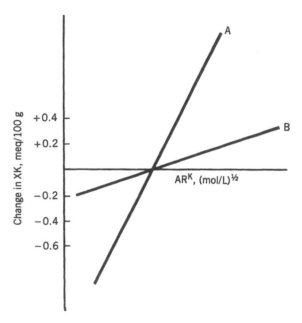

Figure 8.8 Typical quantity-intensity curves. The steeper curve for soil A indicates that, compared to soil B, it has a smaller change in solution K (*I*) with a change in the quantity of exchangeable K (*Q*) and, therefore, a greater potassium buffer capacity (KBC).

8.6.1   Quantity and Intensity Relationships

The quantity and intensity, or Q/I, relationship concept was introduced in 1964 to express the KBC. Similar graphs are produced, except that the ordinate is the change in exchangeable K associated with various K activity ratios, AR^k. To obtain the data for the graph, see Figure 8.8, fixed quantities of solution with the same a_{Ca}^{2+} and a_{Mg}^{2+} are made with varying a_K^+. The solutions are mixed with soil and allowed to stand for a period of time. In effect, this creates soils with differing amounts of XK. Then the solutions and soil are separated and the solutions analyzed for K, Ca, and Mg. The gain or loss of K in the solution represents loss or gain from the soil, respectively. This gain or loss is attributed to a change in the amount of XK and is expressed as change in K in cmol/kg and plotted on the ordinate. The final solution concentrations of K, Ca, and Mg are converted into activities and used to calculate the AR^k of the final solutions and are plotted on the abscissa. Calculation of the AR^k is as follows:

$$AR^k = \frac{a_k^+}{\left(a_{Ca}^{2+} + a_{Mg}^{2+}\right)^{1/2}}$$

The slope of the line in the graph is a measure of the KBC (see Figure 8.8). The steep slope for soil A represents a soil strongly buffered with K, one that has a large capacity to maintain the AR^k by changes in XK (and/or fixed K) during the growing season. Where the line crosses the abscissa at zero change in XK, the soil is at its equilibrium AR^k. The Q/I relationship expresses both the intensity (I) or instant K availability and the subsequent availability (Q).

Extremely weathered soils with kaolinitic and oxidic mineralogy, with little or no fixed K and low exchangeable K, are poorly buffered. Such soils are less able to maintain a satisfactory AR^k as K uptake occurs during the growing season.

8.7 FACTORS AFFECTING UPTAKE OF POTASSIUM

Potassium uptake at the root level is closely related to the concentration gradient between soil and root, rate of K diffusion through soil to root surfaces, and the surface area of roots. Plants grown on different kinds of soils with the same exchangeable K may take up different amounts of K during the growing season and yield differently due to differences in the KBC. The effect of temperature on K uptake is very complex because chemical transformations of K in the soil are temperature dependent, and root growth and other plant processes are affected by temperature. Several other factors that affect K uptake from soils will be discussed.

8.7.1 Soil Moisture

Diffusion of K^+ through soil water to roots is the most limiting step in K uptake during the growing season. As soils dry, K diffusion decreases, making the available K, in a sense, less available for uptake. And as soils dry, there is also less water in which the K can diffuse, and the diffusion paths are more tortuous.

The effects of dry weather on yields and K uptake are variable. If dry weather results in more deeply penetrating roots to grow and forage through a greater volume of soil, yields and total K uptake may increase. On the other hand, if rooting depth is restricted by either too much or too little moisture, the crop may benefit from K fertilization. The results are highly dependent on the forms and amounts of subsoil K and the specific rooting pattern as related to soil and moisture.

8.7.2 Cation Exchange Capacity

As the CEC increases, a given amount of exchangeable K will equilibrate with less K^+ in solution, which is the dilution effect. Fine-textured soils require a higher level of exchangeable K to produce the same available K^+ and yields that coarse-textured soils do. Many states account for variations in CEC due to clay content in making K fertilizer recommendations. For example, the desired pounds of exchangeable K pp2m for general cropping in Ohio is calculated as follows:

$$\text{desired soil test level} = 220 + (5 \times \text{CEC})$$

Soils with higher CEC usually have finer texture and have slower K diffusion rates as compared to coarse-textured soils with the same water content.

The formula does not work well when the CEC is less than 5 $cmol_c$ kg^{-1}, the soils are high in organic matter, or when soils have an excessive amount of XCa.

8.7.3 Exchangeable Cation Suite

An increase in the quantity of any cation increases the solution concentration or activity of that cation when all other factors remain constant. In calcareous soils, the opportunity for Ca^{2+} dissolution is great and a low AR^k may depress K uptake under these conditions. Corn grown on the highly calcareous Harpster soil in Iowa frequently had K-deficiency symptoms when corn grown on nearby noncalcareous Webster soils yielded very well. The yield differences were associated with significant differences in plant composition. The normal and calcareous soils produced different sized plants with nearly identical amounts of K + Ca + Mg in meq/100 g. However, in the normal plants, K made up 60% of the total cations, as compared to only 13% for plants grown on calcareous soil.

8.7.4 Soil pH Influences

In acid soils with exchangeable Al, increasing soil pH results in less exchangeable Al^{3+} and allows K^+ to compete more effectively against Ca^{2+} and Mg^{2+} for adsorption. With more adsorbed K, there is less K in solution. The removal of the interlayer hydroxy-Al of clays greatly increases the potential for K fixation. A pH increase is accompanied by an increase in pH dependent CEC, creating negative charge for adsorption of cations, including adsorption of K^+. The addition of lime adds Ca^{2+} and Mg^+ and decreases the AR^k. All four of these effects reduce solution K^+ or I and the AR^k. By contrast, however, these effects reduce the loss of K from soils by leaching. If liming reduces Al toxicity, plants may grow larger and take up more K. Thus, the relationship between soil pH and K availability is complex, and liming has a highly variable effect on K uptake.

8.7.5 Soil Aeration

Active accumulation of K by roots depends on metabolic energy. When soils become water saturated and respiration is reduced for a lack of O_2, nutrient uptake is reduced. Of the nutrients absorbed in greatest amounts, K has its percentage composition and uptake reduced considerably more by poor aeration than some other nutrients as shown in Table 8.4. The data in Table 8.4 also show how the uptake of one ion, in this case K, and plant growth may be reduced with perhaps normal uptake of other ions and an increase in their percentage composition.

Table 8.4 Effects of Aerating a Saturated Soil on the Percentage of Composition and Uptake of Nutrients by Corn

Treatment	N	P	K	Ca	Mg
Composition, percent					
Aerated	3.33	0.24	1.50	0.63	0.67
Saturated	3.86	0.55	0.65	0.96	0.96
A/S	0.44	0.44	2.31	0.66	0.70
Uptake, grams					
Aerated	0.270	0.026	0.122	0.051	0.054
Saturated	0.181	0.020	0.031	0.045	0.045
A/S	1.490	1.300	3.935	1.130	1.200

Source: Data from Lawton, 1945.

8.7.6 Plant Differences

A difference in the K-exploiting ability of grasses and legumes has received much attention. Legumes are inferior to grasses in their ability to extract soil K. When both are grown together in soils with marginal levels of XK, the legumes may encounter K deficiency while the grasses grow well. Using biotite as a K source in greenhouse experiments, researchers found that wheat can use biotite K and alfalfa cannot. Similar results have been obtained by comparing ryegrass and red clover. The lower root-length density of cotton, as compared to wheat, seems to account for the lower ability of cotton to use soil K.

Differences in K uptake from a given soil by different varieties or cultivars have also been found. It appears that genetic improvement of plants can result in more efficient use of soil K with corresponding yield increases.

8.7.7 Topsoil versus Subsoil Potassium

The importance of subsoil K is related to the level of available subsoil K, relative to the topsoil, on root density, and on the residence time of roots for K uptake. Root development occurs first in the topsoil and root density is typically much greater in the topsoil than in the subsoil. By the time roots have become abundant in the subsoil, the stage of plant development may play a role in uptake. Subsoil K tests are usually lower than topsoil tests, and all of these factors tend to account for greater uptake of topsoil K.

Considerable work on roots and nutrient uptake has been conducted at Purdue University by Barber and his colleagues. They found that for corn (maize) 55% of the roots were in the subsoil and accounted for only 10% of the total K uptake.

A regional field experiment in the north-central region of the U.S. found that including subsoil K tests with topsoil tests did little to improve the correlation between soil tests and K uptake.

The opposite situation holds for some Ultisols of the coastal plains of the southeastern U.S. These soils have very sandy surface soils overlying much finer-textured subsoils; these subsoils retain large amounts of plant available K. Yields of corn and soybeans are significantly affected by subsoil K. In Delaware, sub-surface sampling of such soils is routinely recommended in assessing K fertilizer needs.

8.8 BASIS OF POTASSIUM FERTILIZER RECOMMENDATIONS

The soil solution K concentration and the I value have been used as a measure of K availability. The XK and Q are related to the ability of the soil to maintain K^+ in solution. When experiments are conducted to correlate K uptake and crop yield with I and Q, the best correlations are with Q. This is to be expected when crops remove large amounts of K relative to the amount in solution. Many experiments have shown high correlations between XK and uptake.

To measure K availability, most soil testing labs use an extracting solution of $1\,M$ NH_4OAc adjusted to pH 7. This test is a measure of XK. The amount of K extracted is generally equated to pounds of XK in an acre furrow slice and expressed as pp2m, in the U.S. Experiments are conducted to determine the K fertilizer rate and to develop curves that relate crop yields to fertilizer rate for soils with varying soil test levels. These data are organized into tables and equations for making K fertilizer recommendations for a particular yield goal.

Examples of K fertilizer recommendations equations for potato production on mineral soils in Michigan are as follows:

$$XKs = 175 + 0.50YG - 1.00ST$$

$$XKc = 156 + 0.63YG - 1.25ST$$

where XKs = pounds of K_2O for loamy sands and sandy loams, XKc = pounds of K_2O for loams, clay loams, and clays, YG = yield goal, ST = soil test, exchangeable K, pp2m.

Potatoes expected to yield 400 hundred-weights per acre and soil tests of 100 would have a recommendation of 275 pounds (XKc) and 283 pounds (XKs) of K_2O per acre for the sandy and fine-textured soils, respectively.

For organic soils, the soil test is based on the volume of about a seven-inch plow layer rather than the weight. The fertilizer recommendation for producing the 400-hundred weights of potatoes on organic soils having 100 pounds of K in the plow layer is

$$XKo = 480 - 0.85ST$$

$$XKo = 395 \text{ pounds of } K_2O$$

In highly weathered Oxisols and Ultisols, there may be essentially no fixed potassium to provide for the replacement of exchangeable K. Complimentary ions like Ca and Mg may have an impact toward lowering the AR^k. Thus, it is sometimes recommended that the minimum requirement of exchangeable K be at least 2 to 3% of the cation exchange capacity.

8.9 REPLACEMENT OF POTASSIUM BY SODIUM

Crops with a high uptake potential for Na may respond to additions of Na if K is limiting. Sugar beet, mangold, spinach, swiss chard, and table beet are examples of such crops. Soils that are high in K or have been adequately fertilized with K are unlikely to respond in yield to addition of Na.

REFERENCES

Barber, S.A. 1984. *Soil Nutrient Bioavailability.* John Wiley & Sons, New York.

Barber, S. A., R. D. Munson and W. B. Dancy. 1985. Production, Marketing and Use of Potassium Fertilizers. *Fertilizer Technology and Use.* Soil Sci. Soc. Am., Madison, WI.

Beckett, P. H. T. 1964. Studies on Soil Potassium: I. Confirmation of the Ratio Law: Measurement of Potassium Potential. *J. Soil Sci.* 15:1.

Bertsch, P. M. and G. W. Thomas. 1985. Potassium Status of Temperate Region Soils. In *Potassium in Agriculture.* R. D. Munson, Ed. Am. Soc. Agron., Madison, WI.

Black, C. A. 1968. *Soil-Plant Relationships.* 2nd ed. John Wiley & Sons, New York.

Bouabid, R., M. Badraoui and P. R. Bloom. 1991. Potassium Fixation and Charge Characteristics of Soil Clays. *Soil Sci. Soc. Am. J.* 55:1493–1498.

Cassman, K. G., B. A. Roberts and D. C. Bryant. 1992. Cotton Response to Residual Fertilizer Potassium on Vermiculitic Soil: Organic Matter and Sodium Effects. *Soil Sci. Soc. Am. J.* 56:823–830.

Christenson, D. R., D. D. Warncke, M. L. Vitosh, L. W. Jacobs and J. G. Dahl. 1992. Fertilizer Recommendations for Field Crops in Michigan. *Ext. Bul. E550A.* Michigan State Univ., East Lansing, MI.

Grim, R. E., R. H. Bray and W. F. Bradley. 1937. The Mica in Argillaceous Sediments. *Am. Mineral.* 22:813–829.

Grimes, D. W. 1966. *An Evaluation of the Availability of Potassium in Crop Residues.* Ph.D. thesis, Iowa State University, Ames, IA.

Jackson, J. E. and G. W. Burton. 1958. An Evaluation of Granite Meal as a Source of Potassium for Coastal Bermuda Grass. *Agronomy J.* 50:307–309.

Lawton, K. 1945. The Influence of Soil Aeration on the Growth and Absorption of Nutrients by Corn Plants. *Soil Sci. Soc. Am. Proc.* 10:263–268.

Martin, J. C., R. Overstreet and D. R. Hoagland. 1946. Potassium Fixation in Soils in Replaceable and Nonreplaceable Forms in Relation to Chemical Reactions. *Soil Sci. Soc. Am. Proc.* 10:94–101.

McLean, E. O. 1978. Influence of Clay Content and Clay Composition on Potassium Availability. In *Potassium in Soils and Crops*. Chapter 1. Potash Research Institute of India, New Delhi.

Mengel, K. and E. A. Kirkby. 1980. Potassium in Crop Production. *Advances in Agronomy* 33:59–110. Am. Soc. Agron., Madison, WI.

Mengel, K. and E. A. Kirkby. 1982. *Principles of Plant Nutrition*. International Potash Institute. Worblaufen-Bern, Switzerland.

Mortland, M. M. and K. Lawton. 1961. Relationships Between Particle Size and Potassium Release from Biotite and its Analogues. *Soil Sci. Soc. Am. Proc.* 25:473–476.

Munson, R. D., Ed. 1985. *Potassium in Agriculture*. Am. Soc. Agron., Madison, WI.

Ohio State University Cooperative Extension Service. 1986. *Ohio Agronomy Guide*. Bull. 472. Agronomy Dept., Columbus, OH.

Pierre, W. H. and C. A. Bower. 1943. Potassium Absorption by Plants as Affected by Cationic Relationships. *Soil Sci.* 55:23–34.

Plummer, J. K. 1918. Availability of Potassium in Some Common Soil Forming Minerals. *J. Agr. Res.* 14:297–315.

Pratt, P. F. 1951. Potassium Removal from Iowa Soils by Greenhouse and Laboratory Procedures. *Soil Sci.* 72:107–117.

Rich, C. I. 1964. Effect of Cation Size and pH on Potassium Exchange in Nason Soil. *Soil Sci.* 98:100–105.

Rich, C. I. 1968. Mineralogy of Soil Potassium. In *The Role of Potassium in Agriculture*. V. J. Kilmer, S. F. Younts and N. C. Brady, Eds. Am. Soc. Agron., Madison, WI.

Ritchey, K. D., Ed. 1979. *Potassium Fertility in Oxisols and Ultisols of the Humid Tropics*. International Agric. Bull. 37, Cornell University, Ithaca, NY.

Sadusky, M. C., D. L. Sparks, M. R. Noll and G. J. Hendricks. 1987. Kinetics and Mechanisms of Potassium Release from Sandy Middle Coastal Plains Soils. *Soil Sci. Soc. Am. J.* 51:1460–1465.

Somasiri, S., S. Y. Lee, and P. M. Huang. 1971. Influence of Certain Pedogenic Factors on Potassium Reserves of Selected Canadian Prairie Soils. *Soil Sci. Soc. Am. Proc.* 35:500–505.

Sparks, D. L. 1980. Chemistry of Soil Potassium in Atlantic Coastal Plain Soils: A Review. *Commun. in Soil Sci. and Plant Analysis* 11:435–449.

Tesar, M. B. 1985. Fertilization and Management for a Yield of Ten Tons of Alfalfa Without Irrigation. *Proceedings 1985 Forage and Grassland Conference*. Am. Forage and Grassland Council, March 3-6, Hershey, PA.

Truog, E. and O. J. Attoe. 1945. Exchangeable and Acid-Soluble Potassium as Regards Availability and Reciprocal Relationships. *Soil Sci. Soc. Am. Proc.* 10:81–85.

Zandstra, H. G. and A. F. MacKenzie. 1968. Potassium Exchange Equilibria and Yield Response of Oats, Barley and Corn on Selected Quebec Soils. *Soil Sci. Soc. Am. Proc.* 68:76–79.

CHAPTER 9

Calcium, Magnesium, Sulfur, and Chlorine

Calcium, magnesium, sulfur, and chlorine, along with nitrogen, phosphorus, potassium and sodium, are the eight most-abundant elements taken up by plants from soils. Calcium, Mg, K, and Na are taken up as cations and N, P, S, and Cl are taken up as anions, except for the uptake of some N as NH_4^+. Generally, the total charge of the cations taken up is about equal to that of the anions, except where NH_4^+ is abundant. Calcium, Mg, S, and Cl are generally taken up in lesser amounts than N, P, and K, but in much greater amounts than the micronutrients.

9.1 CALCIUM

Calcium is relatively abundant in the earth's crust, being 3.6% on the weight basis. The amount in soils is highly variable, depending on the content in the parent material and the losses during soil formation. Only rarely is Ca deficient for plants, owing to the large total and available amounts in soils relative to plant requirements.

Once Ca is used in the plant, it is not translocated, so plants must take up Ca during the reproductive stage. As a result, Ca deficiency shows first as a reduction in the growth of meristematic tissues, such as the growing tips and youngest leaves (Figure 9.1).

9.1.1 Forms in Soils

Calcium, similarly to K, is generally abundant in the feldspars of the sand and silt fractions. Calcium occurs in anthorite, $CaAl_2Si_2O_8$, a plagioclase feldspar. Calcium is also very abundant in limestone and in many soils as calcite, $CaCO_3$, and to a lesser extent as Ca phosphates. Mineral weathering releases Ca^{2+} to the soil solution and it is strongly adsorbed by the negative charge of clays and SOM. It is the dominate exchangeable cation in most soils. The XCa establishes an equilibrium with Ca^{2+} in solution and the exchangeable Ca buffers changes in

Figure 9.1 A calcium deficient tomato plant showing dieback of the terminal buds.

solution Ca due to losses by plant uptake and leaching. Mineral weathering releases a continual source of Ca for resupplying the exchangeable and solution Ca.

A soil with a cation exchange capacity (CEC) of 16 cmol$_c$ kg^{-1} that is 67% Ca saturated contains 4,280 pounds of XCa pp2m. This amount, in only the plow layer, is equal to the needs of productive crops for about 86 years (see Table 3.5). By contrast, there is much less XK than XCa and plants need much more K than Ca. In general, there may be ten times more Ca^{2+} than K$^+$ in soil solution, but plants accumulate more K than Ca. Thus, while most soils supply an adequate amount of Ca for crops, K is frequently supplemented with fertilizers.

Calcium is not fixed, in the sense that K is fixed, and is more strongly adsorbed onto humus, relative to clays, than is Mg, K, or Na.

Soil tests for XCa have shown that extractable levels increase with clay content and also increase as the soil drainage changes from well drained to poorly drained (Table 9.1). Thus, the A horizon of poorly drained loamy sand and sand-textured soils contained greater quantities of XCa than the A horizon of well-drained clay and clay loam soils even though the A horizon of well-drained clay and clay loam soils contained nearly three times as much XCa as the well-drained loamy sand and sandy soils. The importance of texture and leaching of Ca is obvious from this data.

Soils that contain free CaCO$_3$ will contain an abundant supply of XCa. In general, these soils are in drier regions where weathering is slow and leaching has not removed soluble and XCa from the soil profile.

Table 9.1 Average Exchangeable Ca Levels of Soil Profiles in Southern Michigan

Dominant profile texture	Profile sample symbol*	Natural drainage class (mg Ca/kg soil)		
		Well	Somewhat poorly	Poorly
Clay and	A	1,725	2,225	3,600
clay loam	B	2,175	2,125	2,950
	C	4,050	3,950	3,825
Loam and	A	1,225	2,175	3,575
sandy Loam	B	1,250	1,850	1,925
	C	1,850	2,250	2,600
Loamy sand	A	625	1,650	2,450
and sand	B	425	750	825
	C	450	1,150	825

* A = plow layer, B = subsoil; and C = parent material.
Source: Robertson, et al., 1976.

9.1.2 Plant Uptake

In most cases, Ca uptake and translocation within the plant are mainly passive processes and greatly influenced by the amount of transpiration. Water flow to the roots of actively transpiring plants generally moves more Ca to root surfaces by mass flow than is needed. This situation produces a Ca concentration gradient where the amount of Ca decreases with increasing distance perpendicularly away from the roots. Where the soil solution is very low in Ca, root interception and diffusion play important roles and Ca depletion occurs adjacent to roots.

The Ca moves toward the apex of the plant in the xylem or transpiration stream. The supply of Ca within plants is reduced by humid, cloudy weather. During fruit development, this condition aggravates Ca deficiency in some crops, like tomato and watermelon, during the fruit development period. Other crops, including peanuts and celery, may show Ca deficiency when grown on soils that supply adequate Ca for many other crops.

Calcium moves downward so slowly in the phloem that root tips are unable to extend into soil that is deficient in Ca. Further, it appears that only the unsuberized or the young active root tips function in Ca absorption.

9.1.3 Causes of Calcium Deficiency

In most soils, the great abundance of XCa and solution Ca, relative to plant needs, makes Ca deficiency for crops rare. Decreasing XCa occurs with increasing soil acidification. Before XCa is likely to be low enough to cause Ca deficiency for cropping, however, lime is usually applied. Liming increases soil pH and increases XCa. In the absence of liming and continued weathering and leaching, some intensively weathered soils will become too low in XCa to supply crop needs. These soils are likely to have a pH less than 5 and cation exchange sites nearly 100% Al saturated. Under these conditions, plant growth is likely to be limited by the deficiency or toxicity of some nutrient other than Ca.

Serpentine soils have a very high imbalance of XMg relative to XCa and may produce Ca deficiency (see Chapter 3).

9.2 MAGNESIUM

Magnesium is a constituent of chlorophyll and is readily transported within plants. When a Mg deficiency occurs, the oldest or lower leaves will show a yellowing. Magnesium deficiency is more common than Ca deficiency, but much less common than K deficiency.

9.2.1 Forms in Soils

The amount of Mg in the earth's crust is lower than Ca and K. Common Mg minerals in soils include biotite, pyroxenes, olivine, and dolomite. These minerals weather quite easily. Magnesium is a structural component of some 2:1 clay minerals, including montmorillonite, and these minerals have much greater weathering resistance. As a result, the amount of Mg in weatherable minerals, relative to the exchangeable Mg, is much less than for Ca and K.

Magnesium ions released in weathering are less strongly adsorbed on clays and soil organic matter (SOM) than Ca and more strongly adsorbed than K. Magnesium commonly makes up 10 to 20% of the exchangeable ions in soils. Magnesium is usually sufficient for cropping where leaching losses are minimal. A soil with a CEC of 16 $cmol_c$ kg^{-1} and 30% Mg saturated would contain 1,152 pp2m, or enough Mg in a typical plow layer to supply crops for 38 years (see Table 3.5).

As soils become more weathered and acid, leaching removes XMg relatively more rapidly than XCa. Consequently Mg deficiencies occur on crops more frequently than Ca deficiencies. Where Mg is deficient for cropping on acid soils, dolomitic lime is recommended for economic reasons. Where high value crops are found to be Mg deficient and a soluble carrier is required, Mg sulfate (epsom salts) may be used in a spray for plants. Generally 10 to 20 pounds of epsom salts in 30 gallons of water sprayed on an acre is adequate. It is recommended that spraying be carried out early in the morning or late in the evening to minimize plant burning. Although dolomite is sufficiently soluble on acid soils, if soil pH is above 6.5 to 7.0, a more soluble carrier of Mg must be used. Epsom salts may be used in this case and the double salt of $K_2SO_4 \cdot MgSO_4$ is also a suitable choice, particularly if K is also needed for the crop.

9.2.2 Plant Uptake

Soil solution Mg is in equilibrium with XMg. The quantity of Mg in solution is about equal to 1 to 10% of the exchangeable Mg. In most minimally and moderately weathered soils, mass flow moves sufficient Mg to root surfaces to satisfy plant needs.

9.2.3   Causes of Magnesium Deficiency

Magnesium deficiencies most often occur on coarse-textured, acid soils that have low clay content, low CEC, and high leaching potential. More specifically, Mg deficiency is likely when the exchangeable Mg is less than 75 pp2m and Mg is less than 3% of the basic exchangeable cations. Potassium represses Mg uptake; thus Mg deficiency is also likely when XK exceeds XMg on a m.e. basis. Grasses with a low content of Mg have a tendency to induce hypomagnesaemia, grass tetany, in dairy cattle. This effect is more often seen in cool, wet spring months when the growth of grass is rapid. In part, this effect may occur because increasing soil solution favors the adsorption of the divalent ions over the monovalent ions. Thus, higher moisture levels in the soil make it more difficult for plants to obtain Mg as compared to K.

9.3 SULFUR

Even though plants use a considerable amount of S, only small quantities must be added to soils to improve their fertility at the present time. This is a paradox, since plants require as much S as P, but P has been considered one of the major fertilizer elements for years. We must consider two factors to understand why only small amounts of S are needed in fertilizers. First, S has been added to soils indirectly for more than 200 years. Gypsum, $CaSO_4 \cdot 2H_2O$, was applied to soils as early as 1768. Superphosphate, developed as a P source and applied to soils after 1850, contained 12% S. One of the first commercial N fertilizers, ammonium sulfate, also contained S. Thus, S was commonly added to the soil along with N and P. A second major factor is that S is added to soils through precipitation; rain furnishes all or part of the S required for growing crops. Industrial development has put more S into the atmosphere, and hence greater quantities of S are added through rainfall.

Modern technology has developed higher analysis N and P fertilizers that do not contain S. Current clean-air standards are reducing industrial emissions of S. These technical advancements would suggest that S levels in soils may be reduced in the future.

9.3.1   Sulfur Content of Crops and Atmospheric Deposition

The S content of crops varies from 45 kg/ha (40 pounds/acre) for alfalfa to 6 kg/ha (5.4 pounds/acre) for rice (Table 9.2). Some generalizations in Table 9.2 are worthy of note. First, grains remove relatively little S compared to a forage crop. Alfalfa, for example, removes five times as much S as wheat and 7.5 times as much as rice. Second, the stover or straw portion of a crop, which is often returned to the soil, contains a significant amount of S. Finally, the particular crop rotation will have a major effect on the soil fertility level of S. Atmospheric deposition of S is quite variable but usually in the range of 3 to 12 kg/ha (2.7 to 10.7 pounds/acre) in the U.S. Thus, a crop rotation that includes a high percentage of alfalfa and forages will remove much more S from the soil each year than is deposited from the atmosphere. On the other hand, a rotation with a high percentage in small grains may nearly balance removal with atmospheric deposition. Near some industrial sites, enough S has been deposited to kill all vegetation.

Table 9.2 Sulfur Content of Agronomic Crops

Crop	Yield	S content, kg/ha
Alfalfa, hay	8 tons	45
Corn, grain	8,000 pounds	17
stover		20
Rice, grain	7,000 pounds	6
straw		8
Soybean, grain	60 bushels	14
Wheat, grain	60 bushels	9

Source: Adapted from Duke and Reisenauer, 1986.

9.3.2   Sulfur Cycles

A simplified S cycle is shown in Figure 9.2. The forms of S added to the atmosphere include SO_2, SO_4^{2-}, H_2S, S^0, dimethyl sulfide (DMS), and other organic sulfides. The sources of the S released to the atmosphere include natural and manufactured materials. Volcanic activity releases significant amounts (65% of total) of S as SO_2, H_2S, and SO_4^{2-}. Soils contribute about one-third of the S. It has been reported that the wetlands release large quantities of DMS to the atmosphere. Oceans are also a source of large amounts of S (24% of total) released to the atmosphere, for the most part as DMS.

Emissions brought about through the activities of human beings are largely associated with power plant operation and automobiles, although other industrial inputs contribute some S. Most of the inputs are as SO_2, H_2S, and SO_4^{2-}.

Once in the atmosphere, all forms can be expected to be oxidized to SO_4^{2-} by reaction with OH^-, O_2, O_3, and other oxidants. But the length of time necessary for this oxidation can be quite variable. The major S forms are oxidized in less than one week, but COS (carbonyl sulfide) is stable in the atmosphere for many years.

Sulfur from the atmosphere enters the soil portion of the S cycle either through dry deposition or through wet deposition. Sulfate is a source of acidity during wet deposition since it commonly associates with H^+ during this deposition. However, wet deposition may also contribute ions with a basic reaction that may offset some of this acidity.

Soluble sulfate, SO_4^{2-}, is the principal inorganic ion in well-aerated soil solutions. If the soil solution contains sulfate S equal to 3 to 5 mg/L, it is considered adequate for plant growth. Plants will utilize sulfate, which reaches

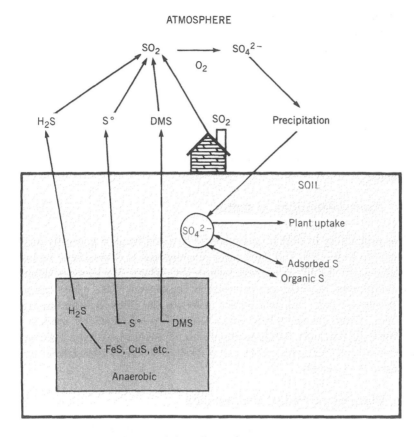

Figure 9.2 A generalized diagram of the sulfur cycle.

the roots by mass flow, diffusion, or both. Generally, mass flow will be the major mechanism if there is greater than 5 mg/L of S in the solution.

Sulfate in solution is also in rapid equilibrium with adsorbed SO_4^{2-} in soils. Hydrous oxides of Al and Fe, as well as exposed Al from edges of clay minerals, are the principal sites of adsorption of sulfate. The adsorption is strongly pH dependent, with the greatest adsorption being found in the very acid soils. Adsorption for a given soil decreases rapidly as pH is increased from 4 to 7. Quantities of adsorbed SO_4^{2-} are usually greater in the subsoil than in the surface soil because of the greater content of clay minerals and Fe and Al hydrous oxides in the subsoil. The possibility of adsorbing SO_4^{2-} on Fe and Al bound to organic matter certainly exists.

More than 90% of the total S in the A horizons of soil exist in the organic form. The N:S ratio in surface soils is relatively constant, with a ratio of about 10:1.3. The nature of this organic S is quite complex. To characterize it, research-ers have fractionated organic S into HI-reducible S, which is S not directly bonded to C. It is usually believed to be ester sulfates (–C–O–S) or related compounds. Most soils will have between 30 and 70% of the organic S as the HI-reducible

S fraction, but soils have been reported that fall considerably out of this range. The more stable organic S fraction is the carbon-bonded S fraction. Usually 10 to 20% of the organic S in A horizons is in this fraction. The remaining fraction is nonreducible S, sometimes referred to as inert or residual organic S.

When soils become poorly aerated (anaerobic), SO_4^{2-} can be reduced to H_2S and other reduced S compounds. This would normally be expected to occur only after other electron acceptors such as Mn^{4+}, NO_3^- and Fe^{3+} have been used. The H_2S may escape to the atmosphere, or form very insoluble precipitates with Fe, Zn, Cu, and other elements. In addition, DMS is one of the major reduced S forms produced in anaerobic soils, and its release to the atmosphere has already been noted.

9.3.3 Plant Responses to Sulfur

The probability of obtaining a response to S fertilization generally increases from eastern to western U.S. Few cases of a response to S have been reported in the northeastern or midwestern states where atmospheric deposition is significant. Sulfur responses, however, are common in the western U.S., where many soils have developed from volcanic parent materials and there is little atmospheric deposition. Irrigation water may often furnish S, so deficiencies are less often found on irrigated lands. Sandy-textured soils are more likely to be deficient than fine-textured soils. Crop responses vary with species, but legumes tend to be more responsive than cereals.

9.3.4 Diagnosing Sulfur Deficiencies

As with other nutrients, S deficiencies have been detected through soil testing and plant analysis. The literature would indicate that plant testing is the more reliable guide, although soil testing does identify potentially S-deficient soils.

A number of extracting solutions have been used to extract soils, all of which aim to remove soluble and adsorbed SO_4^{2-}. A partial, but certainly not complete, list would include $CaCl_2$ (0.01 to 0.1 M), $Ca(H_2PO_4)_2$, 0.25 M HOAc plus 0.15 M NH_4F and 2 N acetic acid containing 500 ppm of P. Although variable with location and crop, the critical level of S is expected to be from 3 to 8 mg/kg. Even when a soil test shows the amount of S to be below the critical level, crops may not respond to applications of S fertilizer because (1) they obtain S from greater depths than were sampled, (2) organic S is mineralized during the growing season, (3) S additions from the atmosphere or water are not accounted for by soil tests, and (4) factors other than S limit yields. In summary, the soil test for sulfate is useful as a guide but will often fail to indicate which soils will respond to applications of S.

Critical values for total S content of plant tissue have been developed. Data compiled in local areas should be consulted for exact critical values, but a few guidelines will be given here. A total S content of plant tissue that exceeds 0.26% may be considered optimum for ryegrass, alfalfa, and clovers. Values that exceed

0.17% may be considered optimum for small grains, and the ear leaf of corn should contain more than 0.12% S to be optimum.

9.3.5 Correcting Sulfur Deficiencies

A number of materials may be used to correct S deficiency, including ammonium sulfate (0-21-0), superphosphate (0-20-0), potassium sulfate, magnesium sulfate, calcium sulfate (gypsum), and elemental S. The recommended rate of S will usually be between 20 and 40 pounds/acre.

9.4 CHLORINE

Chlorine was one of the last elements found to be essential. The plant requirement for Cl is very small and it is very difficult to develop an environment free of Cl for growing plants. The quantities of Cl in the rain and atmosphere are greatly influenced by distance from the sea; the concentration in rain reduces rapidly with distance from the sea.

Although Cl is essential, only a few ppm are required for physiological development of plants. Soils generally contain adequate if not excessive amounts of Cl for plant growth. Not only is Cl added in precipitation, but it is added in fertilizer, particularly KCl and in irrigation water. Organic materials, such as animal manure and sewage sludges, contain Cl.

If soils contain less than 2 mg/kg of Cl, they are considered low. But even those soils may receive adequate Cl from rain to maintain normal plant growth. Chloride is one of the most mobile nutrients in soils and is leached from soils with drainage water.

Salt problems frequently develop in soils that are associated with Cl. Most Cl salts are very soluble; consequently, if a soil is high in Cl salts, the soil solution may contain levels that reduce plant growth.

REFERENCES

Duke, S. H. and H. M. Reisenauer. 1986. Roles and Requirements of Sulfur in Plant Nutrition. In *Sulfur in Agriculture*. M. A. Tabatabai, Ed. Am. Soc. Agron., Madison, WI.

Follett, R. H., L. S. Murphy and R. L. Donahue. 1981. *Fertilizers and Soil Amendments*. Prentice-Hall, Englewood Cliffs, NJ.

Mengel, K. and E. A. Kirby. 1982. *Principles of Plant Nutrition*. International Potash Institute. Worblaufen-Bern, Switzerland.

Nriagu, J. O., D. A. Holdway and R. D. Coker. 1987. Biogenic Sulfur and the Acidity of Rainfall in Remote Areas of Canada. *Science* 237:1189-1192.

Robertson, L. S., D. R. Christenson and D. D. Warncke. 1976. *Essential Secondary Elements: CALCIUM*. Extension Bulletin E-996, CES, MSU.

Tabatabai, M. A., Ed. 1986. *Sulfur in Agriculture*. Am. Soc. Agron., Madison, WI.

CHAPTER 10

Micronutrients

A number of elements that are required by plants in very small quantities are known as micronutrients or *trace* elements. This term usually applies to elements that are contained in plant tissues in amounts less than 100 mg/kg. Although trace elements have been known to affect plant growth for many years, they have been studied intensively since 1950. When other growth factors kept yields at relatively low levels, seldom did micronutrients limit growth and yield. But with the advent of modern fertilizer technology, irrigation, and new varieties, came very high yield potentials. Micronutrient supplies in soils that were adequate for 40, 50, or even 100 bushels of corn per acre were not adequate for yields of 200 or more. Thus, the need to study soil fertility from the micronutrient standpoint became more pressing.

Our ability to study micronutrients has always been closely tied to our analytical capabilities. Although colorimetric methods have existed for many of the micronutrients, they were laborious and often subject to a variety of interferences. When atomic absorption spectrophotometry became common in the early 1960s, it lent itself well to determinations of the trace metals Zn, Cu, Mn, and Fe, as well as other trace metals, such as cadmium and nickel which are potentially toxic. Inductively coupled plasma emission spectrographs gave us the ability to analyze for many elements at the same time, thus greatly reducing the cost of analysis.

10.1 CLASSIFICATION OF MICRONUTRIENTS AND TRACE ELEMENTS

The micronutrients that are essential for plant growth are zinc, copper, iron, manganese, boron, molybdenum, and chlorine. Others such as vanadium, sodium, nickel, cobalt, and silicon may have some function in plant growth. It is obviously very difficult to purify all growth media to the point of proving essentially for a trace element. Consequently, others may be added to this list at a later date. The term *trace element* may be more applicable in the discussion of certain metals that are essential for animal growth or in fact toxic to either plants or animals.

Certain trace elements are essential for animal growth and generally are furnished to the animals (including human beings) by the plant material consumed. Included in this list would be cobalt, chromium, selenium, iodine, and perhaps tin and nickel. Other elements are known to be toxic to plants, animals, or both. These include mercury, lead, and cadmium. For many others, the levels will determine the toxicity or beneficial nature of the element. For example, chromium, which is required in very low levels for animal growth, may enhance plant growth under some conditions. But chromium can be quite toxic to both plants and animals if present in very high levels.

A complete discussion of all trace elements is beyond the scope of this text.* We will focus here on the identification of soils and cropping systems that are likely to be infertile because of micronutrient deficiencies. We will also discuss how to identify and correct the deficiencies. The elements covered will be Zn, Cu, Fe, Mn, B, and Mo.

10.1.1 Essential Micronutrients

Of the six essential micronutrients that we will discuss, four — Cu, Zn, Mn, and Fe — exist as cations in soils and two — B and Mo — exist as anions or uncharged molecular species. The discussion will reflect these differences. A general diagram reflecting the different pathways that micronutrients may take in soils is given in Figure 10.1.

The importance of a particular pathway will depend on the micronutrient and the particular soil. Each may be added to the soil's pool of soluble micronutrients by weathering of minerals, by mineralization of organic matter, or by addition as a soluble salt. Once in the soil as a soluble nutrient, the micronutrient may undergo a number of reactions. Many of the micronutrients will readily precipitate in soils. A given micronutrient may be absorbed by a plant or microorganism. Crop harvest obviously removes micronutrients from the system. Micronutrients are also incorporated into humus, which is formed as plant residues are digested by microorganisms. This process immobilizes micronutrients just as it does other nutrients.

Adsorption of micronutrients, either by soil organic matter or by clay-size inorganic soil components, is an important mechanism of removing micronutrients from the soil solution. Finally, micronutrients may leach from soils. But generally, leaching is a minor component of the mechanisms by which micronutrients are removed from the soil solution.

---

* For those interested, *Applied Soil Trace Elements*, B.E. Davies (Ed.), 1980, and *Micronutrients in Agriculture*, No. 4, J.J. Mortvedt et al. (Eds.), SSSA, Madison, WI, are recommended.

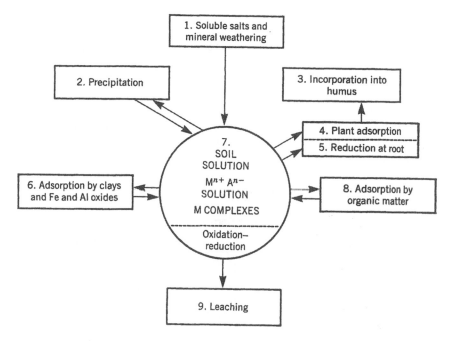

Figure 10.1 The general cycle for micronutrients in soils.

10.2 COPPER AND ZINC

Both Cu and Zn occur in the earth's crust primarily as sulfide minerals. Igneous rocks contain larger amounts of Cu and Zn than do sedimentary rocks, and both elements concentrate more in basalt than in granite. The earth's crust contains on the average 55 ppm Cu and 70 ppm Zn. Soils commonly vary in total elemental content from 2 to 100 ppm Cu and from 10 to 300 ppm Zn.

10.2.1 Copper

Except for copper's occurrence in certain primary minerals, it is bound stronger by adsorption, principally by organic matter, as shown by reaction 8 in Figure 10.1, than by precipitation. Thus, profile distributions of Cu tend to follow the organic matter distribution, with higher concentrations in the surface horizon. These distributions reflect increases in Cu in horizons that have accumulated organic matter (as in spodic horizons). It has been shown that carboxyl and phenol groups are important as the functional groups binding Cu to soil organic matter. Copper is more strongly bound by organic matter than other metals, with the exception of Fe and Al.

The role of organic matter in Cu chemistry is also indicated by analysis of the soil solution. Greater than 99% of the Cu in the soil solution is complexed by organic matter. This complexing is of great importance in maintaining adequate

Cu in solution for plant utilization. Of the inorganic forms of Cu, Cu^{2+} is the major ion form if the solution pH is less than 6.9. The major ion pair is $Cu(OH)_2^0$ if pH is above 6.8. Although $CuOH^+$ does form, it is never significant relative to the other two species.

Deficiencies of Cu are not commonly found in mineral soils. Organic soils (Histosols) containing little ash are more likely to be deficient. When organic soils are deficient, any one of a number of Cu carriers are satisfactory. Some of the common carriers are listed in Table 10.1. The initial application of Cu should be banded at the rate of 6 pounds/acre. Because Cu accumulates in soils, no additional amount need be added for crops that respond little to Cu after 20 pounds/acre has been applied over a period of years. This amount needs to be doubled for highly responsive crops.

Table 10.1 Examples of Copper Carriers

Carrier	Formula	Percent Cu
Basic copper sulfate	$xCuSO_4 \cdot yCu(OH)_2$	13–53
Copper sulfate	$CuSO_4 \cdot H_2O$	35
Cupric ammonium phosphate	$Cu(NH_4)PO_4 \cdot H_2O$	32
Copper EDTA chelate	$Na_2CuEDTA$	13
Copper HEDTA chelate	NaCuHEDTA	9
Copper frit	Frit	40–50
Cupric oxide	CuO	75
Cuprous oxide	Cu_2O	89
Copper chloride	$CuCl_2$	17

Source: Robertson, Warncke, and Knezek, 1981.

If Cu deficiency is found during the growing season, foliar sprays can be used at one-half to one pound Cu per acre, dissolved in 30 gallons of water. Common carriers for this purpose are $CuSO_4$ and CuO. Chelated forms of Cu are well adapted to foliar application. The Cu chelates used at sprays should be applied at the rate of about 35 g Cu/acre dissolved in 30 gallons of water.

10.2.2   Zinc

Much data has accumulated indicating a decrease in solubility of Zn with increasing pH. But precise identification of a solid phase which controls Zn solubility has been difficult. However, Zn uptake by plants declines rapidly as pH increases (Figure 10.2). Of the inorganic Zn species in soil solution, Zn^{2+} is the predominant one for pH less than 7.7, and $ZnOH^+$ is the predominant species for pH between 7.7 and 9.1. The ion pair $Zn(OH)_2^0$ is only important for pHs above 9. Organic matter does complex Zn in soil solution, but the percentage of the Zn that is complexed varies over a considerable range, from 28 to 99 with a mean of 60 for 20 soils, according to one study.

The importance of organic matter in maintaining available Zn in soils is often illustrated by Zn deficiency, which appears in areas where the surface soil has been removed, either by leveling in preparation for irrigation or during the

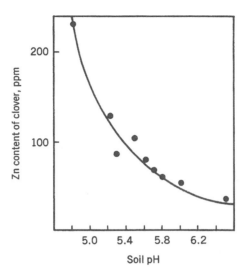

Figure 10.2 The relation between the zinc content of clover and soil pH.

installation of tile drainage lines. Here the effect may also be due to increased pH, since removal of the surface layer very often exposes calcareous B and C horizons.

High levels of P in soils have been known to intensify Zn deficiency in a number of crops. It was particularly noticeable in rotations of navy beans, a crop susceptible to Zn deficiency, followed by sugar beets, a crop that responds to high levels of P fertilization. Data illustrating this are given in Table 10.2.

Table 10.2 Interaction of Phosphorus and Zinc with Navy Bean Yield

Zinc, pounds/acre	Carrier	Yield of navy beans, bushels/acre		
		No extra P	174 extra P	696 extra P
None	—	30.8	20.0	8.8
4.0	Zinc sulfate	39.1	39.8	34.6
4.0	Residual (1 yr)	29.7	24.3	12.1

Source: Judy, Lessman, Rozycka et al., 1964.

The exact cause of the Zn-P antagonism has been difficult to determine. But several factors are important from a soil fertility standpoint. The Zn-P antagonism occurs on calcareous soils and may be related to Fe availability. Moreover, this relationship may not be a soil relationship but one within the plant itself. Applications of Zn will readily overcome the P-induced Zn deficiency. The Zn applications show a considerable residual effect, indicating that total available Zn in the soil is very important in preventing Zn deficiency (Table 10.3).

The experiment was initiated in 1965, but severe drought in the summer of 1965 eliminated yields that year. Residual Zn was very effective in increasing yield, not only through the three years recorded here but for several additional years. The lower yield for the 3 pounds/acre of Zn banded yearly for this site

Table 10.3 Effect of Residual Zinc on Yield of Pea Beans

Zinc, pounds/acre	Carrier	Time of application	Crop yield, bushels/acre 1966	1967
None	—		4.2	0
3.0	ZnSO$_4$	Yearly	17.1	3.7
25	ZnSO$_4$	1965	19.9	19.0
122	Clinker	1965	17.7	18.6

Source: Brinkerhoff et al., 1966, 1967, and Vinande et al., 1968.

shows that this rate of Zn was not adequate for this very Zn-deficient soil. Other data have shown that banding Zn is a satisfactory method of supplying it if sufficient quantity is used.

Carriers for Zn are given in Table 10.4. Inorganic carriers have been satisfactory for correction of Zn deficiency. If the less-soluble forms of Zn such as ZnO are used, they should be finely ground. There is considerable evidence that the less soluble carriers should be broadcast and incorporated into the soil, whereas the soluble carriers, such as ZnSO$_4$ and Zn chelates, should be banded with starter fertilizer at planting time. Rates of 3 to 4 pounds/acre of Zn as inorganic carriers band applied with a starter fertilizer each year are satisfactory. Chelated material may be applied at one-fifth the rate of the inorganic carriers. A single broadcast application of 25 pounds of Zn/acre appears to be adequate for many years.

Table 10.4 Examples of Zinc Carriers

Carrier	Formula	Percent Zn
Zinc sulfate	ZnSO$_4$ · H$_2$O	36
Zinc oxide	ZnO	78–80
Zinc carbonate	ZnCO$_3$	52–56
Zinc EDTA chelate	Na$_2$ZnEDTA	14
Zinc HEDTA chelate	NaZnHEDTA	9
Raplex zinc	ZnPF	10

Source: Robertson and Lucas, 1976.

Considerable success has been obtained by incorporating Zn in starter fertilizers. For example, ZnO incorporated into ammonium polyphosphate becomes soluble and available to plants. Although trace elements, such as Zn, have been found to be beneficial in stabilizing APP liquid fertilizers, we must question the use of high-P starter fertilizer as a carrier of Zn in field situations in which the Zn deficiency is induced by high-P soil levels.

10.3 MANGANESE AND IRON

To a certain extent, Mn and Fe have similar chemistries in soils. Both will exist in more than one oxidation state: Fe^{2+}, Fe^{3+}, Mn^{2+}, Mn^{3+}, Mn^{4+}; consequently, both are affected by drainage conditions of the soils. Both are precipitated as oxides and hydroxides, but Fe forms far less soluble compounds. The total concentration of Mn and Fe in soil solution is increased by complexing with colloidal organic matter.

10.3.1 Manganese

In well-aerated, high-pH soils, Mn is expected to precipitate as MnO_2 and is removed from solution as shown by reaction 2 of Figure 10.1. But as pH decreases, $MnCO_3$ becomes the more stable phase. Here a paradox develops since high CO_2 levels develop in soils when drainage is poor, which decreases the solubility of Mn due to precipitation of $MnCO_3$. But on the other hand, poor aeration favors the reduction of Mn^{4+} to Mn^{2+} and reduced Mn compounds are more soluble.

Soluble Mn is thought to be in the form of Mn^{2+} but it has been shown that 80 to 90% of the Mn in soil solution is complexed with organic matter. Steam sterilization reportedly makes Mn more soluble because it reduces and hydrates Mn compounds. It is true that steam sterilization releases Mn, but the reason appears to be that steam alters the functional groups on organic matter. We do not know whether altering the functional groups releases more colloidal organic matter to soil solution, which is capable of chelating Mn, or whether it reduces the soils ability to adsorb Mn (Table 10.5).

Table 10.5 Effect of Steam Treatment on Change in Mn Fractions of a Houghton Muck

Mn added ppm	Time before extraction days	Steam treatment	Change in Mn extracted, ppm		
			Water	Exchangeable	Easily reducible
0	—	5 hours	3.0	18	−8.5
800	0	none	49	324	107
800	35	none	0	0.4	216
800	35	5 hours before extraction	50	342	101
800	35	5 hours before Mn addition	28	299	138

Source: Calculated from data of Kozakiewicz and Ellis as given in *Applied Soil Trace Elements,* 1980, Brian E. Davies, Ed., John Wiley & Sons, New York.

The availability of Mn in the field has always been difficult to predict. A number of reasons may account for this. Since Mn solubility is related to oxidation-reduction reactions in the soil, the availability of Mn is closely related to weather. Cool temperatures may slow down the mineralization of organic Mn. On the other hand, cool temperatures associated with high levels of rainfall in early spring may keep more Mn available due to reduction of Mn oxides.

There is an interaction between Mn and Fe. High levels of available Fe in organic soils or high levels of organic matter in sands may lead to Mn deficiency because of the high ratio of Fe to Mn within the plant. This ratio is particularly important since certain chelated Mn carriers will actually make the situation worse rather than correcting Mn deficiency. Data in Table 10.6 shows that the application of as little as one pound Mn per acre as MnEDTA actually reduced the yield of soybeans by about 50%. Similar data were observed for onions. The reason is that the soil had high levels of available Fe and low levels of available Mn. The Mn added as the chelate readily dissociated and was apparently rendered unavailable, leaving the chelate to complex more Fe, thereby increasing the available Fe.

Table 10.6 Interactions Between Manganese Carriers, Soil pH and Yield of Soybeans

| Mn treatments | | Planting time fertilizer | Yield of soybeans, bushels/acre | |
Mn, pounds/acre	Carrier		pH 6.4	pH 7.5
0	None	Acid	19.2	22.1
10	MnSO$_4$	Acid	30.8	30.1
1	MnEDTA	Acid	17.2	24.1
0	None	Neutral	19.2	24.0
10	MnSO$_4$	Neutral	23.7	30.9
1	MnEDTA	Neutral	10.9	13.4

Source: Rumpel et al., 1967.

A number of Mn carriers may be used to correct Mn deficiency as shown in Table 10.7. Manganese sulfate has been the most satisfactory material for most situations. The inorganic carriers MnO and Mn frit are not water-soluble carriers and must be finely ground to be satisfactory; finer than 100-mesh is essential and finer than 300-mesh is desirable. MnEDTA is not satisfactory for organic soils and sands high in organic matter that may contain a high ratio of available Fe to Mn.

Table 10.7 Examples of Manganese Carriers

Carrier	Formula	Percent Mn
Manganese sulfate	MnSO$_4$ · 3H$_2$O	26–28
Manganous oxide	MnO	41–68
Manganese frit	Frit	35
Manganese EDTA chelate	MnEDTA	12
Various other organic Manganese complexes	Mn-organic	5–12

Source: Robertson and Lucas, 1976.

The recommended rate of Mn application when Mn deficiency is suspected varies with soil pH and mineral content. Generally, if soil pH is above 6.5, from 4 to 8 pounds/acre of Mn is recommended for mineral soils. If pH is from 6.0 to 6.5, from 4 to 6 pounds of Mn is adequate. In all cases the Mn should be band-placed at planting time for best results. For organic soils, if soil pH is above 6.4 an application of from 4 to 16 pounds of Mn/acre is recommended, depending upon the severity of the deficiency. If the pH is from 5.8 to 6.4, then 4 to 12 pounds of Mn per acre is recommended. Foliar sprays with Mn may also be used if they are compatible with other spraying programs or if the Mn deficiency appears after the crop is planted.

10.3.2 Iron

Few, if any, soils are deficient in total Fe since the total Fe content of soils varies from 1,000 to 10,000 ppm. But the solubility of Fe in soils may be limited by the low solubility of Fe hydroxides and Fe oxides in the pH range in which crops are grown.

Soil conditions which lead to Fe deficiency in plants include pHs above 7.0, low moisture content, and low organic matter content. These conditions are encountered in the more arid western states of the U.S. When Fe deficiency is seen in other areas, for example in the north-central region of the U.S., it is normally associated with removal of the surface soil and exposure of calcareous subsoils. In these areas, iron-inefficient species, pin oaks for example, may exhibit severe Fe deficiency.

Because of the very limited quantity of Fe^{3+} ions in the soil solution of calcareous soils, it is obvious that organic matter must play a significant role in keeping Fe in soil solution by forming very strong Fe-organic matter complexes. Plants then obtain Fe from these complexes by reducing the Fe^{3+} to Fe^{2+} at the root surface, as shown in reaction 5 of Figure 10.1, and thereby freeing the Fe from the organic complex.

The rate of reduction of ferric to ferrous iron has been found to be greatly increased when a plant is under Fe stress. Staining techniques have been developed to reveal the location of active Fe-reducing sites (Bell et al., 1988). Figure 10.3 shows that the active sites in tomato plants are on the younger root hairs located either on lateral or near the tip of the primary root. As is evident, the plants that are grown without Fe and with P added show much more of the staining. A high magnification of lateral root hairs confirms that the site of the reduction is on the root hair and not on the epidermal cells between the root hairs (Figure 10.4). The dark staining indicates intense activity on these root hairs of an Fe-stressed plant. The plant's ability to respond to Fe-deficient conditions by becoming able to reduce Fe^{3+} at the root surface is remarkable and undoubtedly accounts for the rather low number of soils that produce plants with an Fe deficiency in the field, even under calcareous conditions.

Correcting an Fe deficiency is very difficult because it is caused by chemical conditions within the soil and not from low total Fe content. If soluble Fe is added to the soil, it is very quickly precipitated and then is not available to plants. Consequently, treatment of the deficiency must be limited to acidification of the soil, thereby solubilizing some of the Fe present. It is also possible to add Fe as a chelate that is so stable that it does not dissociate in the soil. Spraying plants with a soluble Fe source is another possible Fe treatment.

The chelates FeEDDHA and FeHEDTA are satisfactory under many soil conditions. Soluble carriers are useful for spraying Fe-deficient plants. Some information about these carriers is given in Table 10.8.

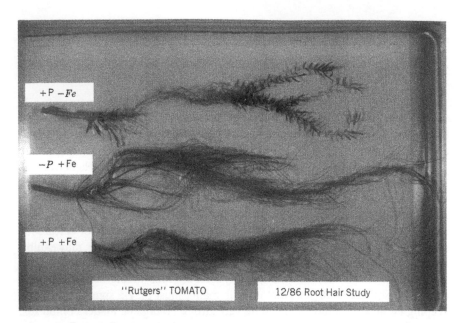

Figure 10.3 Tomato roots showing the location of iron reduction sites in response to iron stress in plants. Notice the staining of the lateral roots (darker areas) on the −Fe plants. (Photograph courtesy P. F. Bell and R. L. Chaney. University of Maryland, the USDA, Beltsville.)

Figure 10.4 A close-up of iron-stressed lateral root hairs. Notice the staining on the root hairs but not the epidermal cells between the root hairs. (Photograph courtesy P. F. Bell and R. L. Chaney. University of Maryland, the USDA, Beltsville.)

Table 10.8 Examples of Iron Carriers

Carrier	Formula	Percent Fe
Ferrous ammonium phosphate	$Fe(NH_4)PO_4 \cdot H_2O$	29
Ferrous sulfate	$Fe_2(SO_4)_3 \cdot 4H_2O$	23
Fe chelates	FeEDTA	9–12
	FeHEDTA	5–9
	FeEDDHA	6

Source: Robertson, Warncke, and Knezek, 1981.

10.4 BORON AND MOLYBDENUM

Boron and molybdenum are unique micronutrients because they exist in the soil as either anions or uncharged molecules. Because they take these two forms, their chemistry in the soil and factors that affect their availability are quite different from those of the other micronutrients. Both B and Mo, however, are much more strongly adsorbed by soils than other anions, such as Cl⁻ or NO_3^-.

10.4.1 Boron

Boron exists as undissociated H_3BO_3 or as the anion $B(OH)_4^-$ in soils and in soil solution. Either form should be mobile in the soil solution. But both forms are adsorbed strongly by either Fe or Al hydroxides. Freshly precipitated Fe or Al hydroxides are known to adsorb much more B than the hydroxides that have aged and crystallized. The bonding is through the hydroxyls at the surface of the precipitated Fe or Al hydroxides. Hence, the more crystalline the hydroxides, the fewer the number of exposed hydroxals per unit weight of hydroxide. Micaceous clay minerals also adsorb B. Magnesium hydroxides and coatings of other minerals that contain Fe, Al, or Mg hydroxides will adsorb B.

Boron deficiency is often accentuated when soil contains little moisture. Consequently, symptoms of B deficiency will very often be observed during dry periods, but after the soils are brought back to field capacity by rain, the new growth will not show B deficiency.

In addition to deficiencies of B, toxicities must also be considered. Boron may be added in irrigation waters or in sludges and wastewaters. Toxicities are very crop dependent; thus, beans may show severe toxicity, whereas sugar beets, under the same conditions, will not show toxicity. A number of B carriers are shown in Table 10.9.

10.4.2 Molybdenum

Concentrations of Mo in soils are very low. Its availability is generally limited by adsorption of MoO_4^{2-} rather than by precipitation. Hydrous Fe oxides and hydroxides are known to adsorb Mo strongly, which undoubtedly explains why Mo deficiencies most often occur on very acid soils. In fact, liming alone will

Table 10.9 Examples of Boron Carriers

Carrier	Formula	Percent B
Borax	$Na_2B_4O_7 \cdot 10H_2O$	11
Boric acid	H_3BO_3	17
Sodium pentaborate	$Na_2B_{10}O_{16} \cdot 10H_2O$	18
Sodium tetraborate	$Na_2B_4O_7$	20
Sodium tetraborate pentahydrate	$Na_2B_4O_7 \cdot 5H_2O$	14
Boron frits	Frits	10–17

Source: Robertson, Lucas, and Christenson, 1981.

generally correct Mo deficiency. But in certain areas of the world where lime is not available, a few ounces per acre of Mo will substitute for many tons of lime.

Carriers for Mo are listed in Table 10.10. Usually about one-eight to one-quarter pound/acre of Mo is adequate to correct Mo deficiency. Since this small quantity of Mo is very difficult to spread evenly, it is better to incorporate Mo with other fertilizer or to use it as a seed coating.

Table 10.10 Examples of Molybdenum Carriers

Carrier	Formula	Mo, percent
Ammonium molybdate	$(NH_4)_6Mo_7O_{24} \cdot 2H_2O$	54
Sodium molybdate	$Na_2MoO_4 \cdot 2H_2O$	39
Molybdenum trioxide	MoO_3	66

Source: Robertson, Warncke, and Knezek, 1981.

10.5 TRACE ELEMENTS THAT MAY BE TOXIC

A number of the micronutrients and other trace element may be toxic to plants, to animals, or to both. There are soils that naturally have large quantities of what are usually trace elements, for example Se. These soils may pose a threat because of this naturally high level of a particular trace element. The number of acres affected naturally is rather few, but in the case of Se, many soils in the western U.S. naturally contain large quantities.

Manufacturing processes have left waste materials that have high levels of many elements including Zn, Cu, Cd, Ni, Cr, Hg, Pb, and others. A few of these will be discussed to illustrate problems and questions which may come up when sludge is utilized as part of a soil fertility program.

Either municipal sludge or sludge generated as a waste product of industry may be available for disposal on land or for recycling on agricultural land to obtain benefit from one or more components of the waste material. The question that arises is whether we can safely use the waste material and obtain a benefit from it. Several rules should be followed before using a waste product. It must be analyzed to determine its exact composition. Normally N in sludges will be similar to N in manure in availability and may be used to advantage. Phosphorus and other nutrients, if needed by the soil, may also be beneficial. Occasionally micronutrients such as Zn may be needed, and sludges can furnish those.

Sludges or waste material will contain other trace metals that are not needed by crops and are potentially toxic. Here the question is how much can be applied to soils before they become a problem. Guides for soil loading have been defined and then redefined as more research data becomes available. Early guides for soil loading based on soil CEC were issued. These and more recent guides are given in Table 10.11. Loading with Pb is much more restricted under the newer guidelines. The other metals are much less restricted. State and local regulations may be more restrictive than the EPA guidelines.

Table 10.11 Total Amount of Metals Suggested for Agricultural Soils Treated with Sewage Sludge

Metal	Maximum amount of metal, pounds/acre Soil Cation Exchange Capacity (me/100g)			503 Cumulative loading rate pounds/acre
	less than 5	5–15	>15	
Lead	500	1,000	2,000	270
Zinc	250	500	1,000	2,520
Copper	125	250	500	1,350
Nickel	50	100	200	380
Cadmium	5	5	5	35

Source: Jacobs, 1981. 503.13 CFR amendments to title 40, Feb. 19, l993.

Certain metals, such as Zn, are not particularly toxic to plants or animals. Consequently, the soil can tolerate a considerable loading with little problem. For example, 503 CFR allows 2,520 pounds/acre cumulative loading of Zn. A heavy loading of Zn and other metals should be accompanied with pH control to maintain pH above 6.5, which will minimize the solubility of the metal. It should also be noted that certain crops may be susceptible to high soil Zn levels.

Other metals, Cd in particular, must be restricted because they are potentially toxic. Cd is a known carcinogen. Furthermore, it readily moves from the soil to the plant root and is easily absorbed by plants. Thus, even relatively low levels in soils becomes a threat to people consuming the food grown on that soil. Very careful control of applications of Cd are essential. The Environmental Protection Agency initially took a very conservative view and restricted Cd to 4.5 pounds/acre total loading. After extensive research, this has been extended to 35 pounds/acre. Lead, although very toxic if consumed, is strongly held by soils and does not solubilize and move readily plant into roots. This means that Pb and Hg may be applied to soils in larger quantities than Cd.

Chromium presents an interesting case. The form used in industry is generally Cr^{+6}. In this form, Cr combines with oxygen and exists as $Cr(OH)_4^{2-}$, which is toxic to both plants and animals. Since it is an anion, it is mobile in the soil. If $Cr(OH)_4^{2-}$ is applied to the soil, it can move out of the rooting zone and to the groundwater. But soils have the ability to reduce Cr^{6+} to Cr^{3+}. This reduction is favored by a high content of organic matter and a low soil pH. Once reduced, the Cr is held as an exchangeable ion and also precipitated as the very insoluble Cr_2O_3. In this form, it is not available to plants.

10.6 METHODS OF EVALUATING SOIL FERTILITY FOR MICRONUTRIENTS

The soil fertility for micronutrients can be evaluated by soil sampling, extraction of an "available" or labile fraction, and analyzing the extract. Also, plant analysis is well adapted to evaluating the micronutrient availability in soils.

10.6.1 Soil Testing for Micronutrients

Deficiencies of micronutrients are related to plant species, climate, and soil chemical properties, such as pH and organic matter content, in ways that have made it very difficult to develop a single soil extractant for all micronutrients. Perhaps the most universally used extractant for micronutrients and other nonessential trace metals is the DTPA test developed by Norvel and Lindsay (1978). The extract consists of 0.005 M DTPA (diethylene triamine pentaacetic acid), 0.1 M triethanolamine, and 0.01 M $CaCl_2$ with a pH of 7.3. The DTPA is a strong complexing agent for heavy metals, particularly Zn^{2+} and Cu^{2+}. Although this test has been widely shown to correlate well with available Zn in soils, it has been less successful in measuring other available micronutrients.

Some extractants that have been used successfully for micronutrients are listed in Table 10.12. For best evaluation and recommendations, they must be coupled with soil pH, soil type and local yield correlation studies.

Table 10.12 Extractants for Micronutrients for Soil Testing

Trace	Soil Extractant
Zn	DTPA
	0.1 N HCl
Mo	Calcium phosphate
Cu	0.1 N HCl
B	Hot water
Mn	Phosphoric acid
	0.1 N HCl
Fe	DTPA

10.6.2 Tissue Testing for Micronutrients

Micronutrient deficiencies may be diagnosed by analyzing the plant tissue. The methods will vary widely with the particular crop and growing conditions. Generally, a certain plant portion is selected (e.g., ear leaf for corn) and a certain stage of plant growth is used. The tissue is collected and analyzed, and the values obtained are compared with values obtained from high-yielding plants. Some data are given in Table 10.13 as general guidelines for the micronutrients.

Table 10.13 General Guidelines for Evaluating Micronutrient Content in Mature Leaves

Micronutrient	Concentration in leaves, mg/kg	
	Deficient	Sufficient
B	<15	20–200
Cu	<4	5–20
Zn	<20	25–150
Fe	<50	50–250
Mn	<20	20–500
Mo	<0.1	0.5–5 (?)

Source: Jones, 1972.

10.7 MICRONUTRIENT DEFICIENCY SYMPTOMS

Deficiency symptoms have been used for a long time to identify deficiencies in the field. This has particularly been true for trace element deficiencies. But a number of factors make this practice difficult and at times less than desirable. It must be recognized that when a nutrient deficiency symptom appears on a plant, there has already been a loss in yield for that year's crop. Correcting the deficiency may be important for that year's crop, even though a yield reduction is expected, and it is important to identify what deficiency to expect in subsequent years. Trace element deficiencies are often related to climatic growing conditions, and these may change from year to year. If growing conditions are not taken into consideration, conclusions about yield responses to applied trace elements can become erratic.

Deficiencies in plants most often are manifested in growth irregularities, so that distinguishing between two or more deficiencies may be difficult. In addition, other factors that affect growth, such as weather, chemical damage, and pest damage to crops, may also give similar symptoms. For these reasons, it is important to obtain all possible information from the grower before attempting to diagnose a deficiency by plant symptoms. Once a tentative identification is made, it should be verified by treatment with the element in which the plant is assumed to be deficient. Table 10.14 has been given to help summarize factors which may help to identify a particular deficiency symptom.

Since many deficiency symptoms are similar, identifying where the deficiency occurs may be very important. Some trace elements, such as Mo, are relatively mobile in plants. When the element becomes deficient in the soil, the Mo in the plant is translocated from the old to the new tissue and the deficiency symptoms appear first on the old tissue. On the other hand, if the element, Fe for example, is not mobile in the plant, then the deficiency appears first on the new growth of the plant.

Table 10.14 Guide to Nutrient Deficiency Symptoms for Micronutrients

Trace element	Plant mobility	Deficiency symptoms	Associated growth changes
Zn	Partially mobile	Light interveinal tissue in older leaves. Abnormal shaped leaves. Shortened internodes of broadleaf crops.	Iron accumulates at nodes. Delayed maturity.
Mo	Yes	General yellowing, with some mottling and cupping of older leaves.	Stunted growth.
Cu	No	Youngest leaves become "olive" and stunted.	Iron accumulates in nodes.
B	No	Unusual brittleness of stems, cracking of stems, thickening, curling, and chlorosis of leaves.	Slow growth. Death of terminal tissue.
Fe	No	Very light yellow to white on new growth. Veins are green.	New growth is severely retarded.

10.7.1 Zinc

Zinc deficiency is perhaps one of the easiest to recognize under field conditions. It most often occurs on calcareous soils and soils that have high levels of phosphate, either as residual or as a current P application. In general, the Zn deficiency appears early in the growing season and is caused by either cool weather or by the restricted rooting zone of plants. Zinc deficiency is also commonly associated with particular crops and, in fact, with particular varieties. For example, when grown on the same soil and in the same rotation, navy beans may show a severe Zn deficiency when sugar beets show no Zn deficiency and do not give a yield response to Zn. Sanilac beans may show a severe deficiency of Zn when Saginaw beans grown side by side give relatively little response to Zn. The same observation can be made about toxicity. Sanilac beans are much more susceptible to Zn toxicity than Saginaw beans if Zn levels are excessive in soils. It is common to find this differential susceptibility to Zn deficiency in different varieties of navy bean, corn, and probably other crops. Breeding to reduce plant susceptibility to Zn deficiency has been successful.

The outward appearance of Zn deficiency is light interveinal tissue. For dicot crops such as navy beans, this disorder appears first in the older leaves and to a lesser degree in younger leaves. The light interveinal tissue takes on the appearance of striping in corn and does not usually affect the older leaves, but rather appears first on leaves of intermediate age. In corn and sorghum, the plants will have shortened internodes (Figure 10.5) and darkened nodal tissue, indicating accumulation of Fe at the nodes. As Zn deficiency develops, the shapes of leaves often become abnormal, particularly in crops such as beans and fruit trees (Figure 10.6). This symptom is very useful in separating Zn deficiency from Mn deficiency since Mn deficiency does not result is abnormal shaped leaves. Delayed

Figure 10.5 Zinc-deficient corn (left) showing shortened internodes.

Figure 10.6 Zinc-deficient bean leaf (right) showing light interveinal tissue and an abnormal
leaf shape.

maturity is also characteristic of Zn deficiency. The field results with navy beans
in Michigan revealed a Zn deficiency so severe that there was essentially no yield.

10.7.2 Molybdenum

The deficiency of Mo may affect legume and nonlegume plants quite differently. Mo is required by *Rhizobia* for fixation of N_2. Consequently, a Mo

deficiency in a legume plant may be manifested as a N deficiency. It then appears as light color and stunting in plants.

Field symptoms of Mo deficiency in the U.S. have most often been observed on vegetable crops. The youngest leaves are most often affected. They become mottled and their leaf margins are narrow. The leaves will elongate abnormally and in cauliflower they are often twisted, leading to the common term *whiptail* applied to the deficiency symptom. Sometimes the leaves may take on a cupping appearance.

10.7.3 Manganese

Manganese deficiency often appears as interveinal chlorosis. Unlike what happens with Zn deficiency, leaves appear normal except for color and show no abnormal leaf shape. If Mn deficiency develops when the plant is very young, the deficiency may be very uniform in both young and older leaves. If the deficiency develops after the plant is much larger, however, it will be much more prevalent on the young leaves. This distribution is uneven, because Mn is only slightly mobile in the plant. In Mn-deficient bean plants, the tissue between the veins becomes increasingly lighter in color. But the veins remain dark green, making this Mn deficiency easy to distinguish from N deficiency (Figure 10.7).

Small grains quite often show Mn deficiency. Generally, a grey oval-shaped spot develops on the edge of a new leaf. The grey spot will enlarge until it covers much of the leaf and takes on a yellow appearance. The tip of the leaf will remain green during this process. Manganese deficiency in corn and grain sorghum appears as interveinal chlorosis, usually on the youngest leaves. It may be similar to Fe deficiency, but Mn deficiency usually appears on soils with high organic matter content, whereas Fe deficiency seldom occurs on these soils.

On certain vegetable crops, such as broccoli, Mn deficiency causes the leaf surface to lose its waxy coating. This is quite apparent when comparing deficient plants with plants sufficient in Mn.

10.7.4 Copper

Because Cu is not translocated in the plant, the deficiency symptoms appear on the new growth. For small grains and corn, the leaves appear olive or yellowish green in color, and often leaves fail to unroll as they emerge. Often the leaf tips will appear as though the plants have been frost damaged, and there will be some flags. A *flag* is a wilted or dead leaf or a branch with such leaves on an otherwise healthy-appearing plant.

10.7.5 Boron

Boron deficiency symptoms are seldom observed except for sensitive crops, such as legumes, sugar beets, and some vegetable crops. In alfalfa, the deficiency is shown by yellowing of the leaves and by restriction of the terminal growth.

Figure 10.7 Manganese-deficient bean plant grown in a greenhouse.

This restriction gives very short internodes and offers a method of separating B deficiency from insect damage, which may give similar visual symptoms, but with normal node lengths. Boron-deficient sugar beets develop cross-checked petioles and misshapen leaves. The leaf blades grow uneven on two sides of the plant and more in the horizontal direction than in the vertical direction. When the deficiency is severe, the terminal growth or apical meristem tissue dies and the roots may develop heart rot.

10.7.6 Iron

Iron is the least mobile of the micronutrients in plants. When the deficiency appears, it is on the new growth and may be very severe. The plant leaves will first appear yellow interveinally with green veins. But when the deficiency is severe, the entire area may appear white (Figure 10.8). It can be demonstrated that Fe is very immobile by placing a local spot of Fe solution on the surface of a deficient leaf and observing that the leaf will develop a green color in that area only.

Figure 10.8 Iron-deficient bean plants showing severe chlorosis on the new growth.

REFERENCES

Brinkerhoff, F., B. Ellis, J. Davis and J. Melton. 1966. Field and laboratory studies with zinc fertilization of pea beans and corn in 1965. *Quart. Bull. of Mich. Agr. Exp. Sta.* 48:344–356.

Brinkerhoff, F., B. Ellis, J. Davis and J. Melton. 1967. Field and laboratory studies with zinc fertilization of pea beans and corn in 1966. *Quart. Bull. of Mich. Agr. Exp. Sta.* 49:1–14.

Chaney, R. L., J. C. Brown and L. O. Tiffin. 1972. Obligatory reduction of ferric chelates in iron uptake by soybeans. *Plant Physiol.* 50:208–213.

Couch, Elton L. and R. E. Grim. 1968. Boron fixation by illites. *Clays Clay Miner.* 16:249–256.

Davies, Brian E., Ed. 1980. *Applied Soil Trace Elements.* John Wiley & Sons, New York.

Follett, Roy H., L. S. Murphy and R. L. Donahue. 1981. *Fertilizers and Soil Amendments.* Prentice-Hall, Englewood Cliffs, NJ.

Hodgson, J. F., W. L. Lindsay and J. F. Trierweiler. 1966. Micronutrient cation complexing in soil solution: II. Complexing of zinc and copper in displaced solution from calcareous soils. *Soil Sci. Soc. Am. Proc.* 30:723–726.

Geering, H. R., J. F. Hodgson and C. Sdano. 1969. Micronutrient cation complexes in soil solution: IV. The chemical state of manganese in soil solution. *Soil Sci. Soc. Am. Proc.* 33:81–85.

Jacobs, Lee W. 1981. Agricultural application of sewage sludge. In *SLUDGE and its Ultimate Disposal.* Borchardt, Redman, Jones and Spregue, Eds. Ann Arbor Science Publishers The Butterworth Group. Ann Arbor, MI.

Jones, J. Benton, Jr. 1972. Plant Tissue Analysis for Micronutrients. In *Micronutrients in Agriculture.* J. J. Mortvedt, P. M. Giordano and W. L. Lindsay, Eds. Soil Sci. Soc. Am., Madison, WI.

Judy, W., G. Lessman, T. Rozycka, L. Robertson and B. Ellis. 1964. Field and laboratory studies with zinc fertilization of pea beans. *Quart. Bull. of Mich. Agr. Exp. Sta.* 46:386–400.

Knezek, B. D. and H. Greinert. 1971. Influence of soil iron and manganese chelate interactions upon the iron and manganese nutrition of bean plants (*Phaseolus vulgaris L.*). *Agron. J.* 63:617–619.

Lindsay, W. L. and W. A. Norvell. 1978. Development of a DTPA soil test for zinc, iron, manganese and copper. *Soil Sci. Soc. Am. J.* 42:421–428.

Mortvedt, J. J., P. M. Giordano and W. L. Lindsay, Eds. 1972. *Micronutrients in Agriculture.* Soil Sci. Soc. Am., Madison, WI.

Robertson, L. S., R. E. Lucas and D. R. Christenson. 1981. BORON: An Essential Plant Micronutrient. *Coop. Ext. Ser. Ext. Bull. E1037.* Michigan State University.

Robertson, L. S., D. D. Warncke and B. D. Knezek. 1981. COPPER: An Essential Plant Micronutrient. *Coop. Ext. Ser. Ext. Bull. E-1519.* Michigan State University.

Robertson, L. S., D. D. Warncke and B. D. Knezek. 1981. IRON: An Essential Plant Micronutrient. *Coop. Ext. Ser. Ext. Bull. E-1520.* Michigan State University.

Robertson, L. S., D. D. Warncke and B. D. Knezek. 1981. MOLYBDENUM: An Essential Plant Micronutrient. *Coop. Ext. Ser. Ext. Bull. E-1518.* Michigan State University.

Robertson, L. S. and R. Lucas. 1975. Essential Micronutrients: ZINC. *Coop. Ext. Ser. Ext. Bull. E-1012.* Michigan State University.

Robertson, L. S. and R. Lucas. 1976. Essential Micronutrients: MANGANESE. *Coop. Ext. Ser. Ext. Bull. E-1031.* Michigan State University.

Rumpel, J., A. Kozakiewica, B. Ellis, G. Lessman and J. Davis. 1967. Field and laboratory studies with manganese fertilization of soybeans and onions. *Quart. Bull. of Mich. Agr. Exp. Sta.* 50:4–11.

Vinande, R., B. Knezek, J. Davis, E. Doll and J. Melton. 1968. Field and laboratory studies with zinc and iron fertilization of pea beans, corn and potatoes in 1967. *Quart. Bull. Mich. Agr. Exp. Sta.* 50:625–636.

CHAPTER 11

Nitrogen, Phosphorus, and Potassium Fertilizers

The fertilizer industry developed in stages with phosphate materials in the 1840s, potassium materials in the 1870s, and the nitrogen materials about 1900. More recently, micronutrient fertilizers have been developed.

Overall soil fertility in the U.S. has been increasing, and in many instances the application of only one nutrient is desirable. Improvements and greater use of soil tests have helped identify these situations. As a result, the use of single-nutrient carriers for direct application to the soil, relative to use of mixtures, has been increasing. Actually, the application of both kinds of fertilizers have been increasing. In 1976, the direct application of single carrier fertilizers surpassed the use of mixtures or mixed fertilizers.

11.1 NITROGEN FERTILIZERS

Manure was the dominant fertilizer for thousands of years in areas where animals were used for power and food. Most of the nutrients in feed appear in the manure, and if the manure is efficiently managed, it can be very effective in the maintenance of soil fertility. Peruvian guano was the first fertilizer imported into the U.S. in 1824. This organic fertilizer consisted of the excreta and remains of birds whose primary diet consisted of fish. Guano contains about 13% N, which is mostly organic. In 1830, $NaNO_3$ was imported from Chile. This inorganic material of natural origin contains about 16% N. These materials were used in the southern U.S. largely on specialty crops, such as cotton and tobacco. In the 19th century, the byproduct NH_3 produced from the coking of coal used in steel manufacturing was neutralized with H_2SO_4 to produce $(NH_4)_2SO_4$.

11.1.1 The Major Nitrogen Carriers

Economics drives the production of N fertilizers toward the material or carrier that supplies N at the least cost. This carrier is anhydrous ammonia, NH_3, which contains 82% N. Its cost advantage is due to both low costs in production and in transportation, for NH_3 is transported by pipeline. Two large pipelines transport NH_3 from plants in Louisiana, Texas, and Arkansas to terminals as far west as Aurora, Nebraska, as far north as Garner, Iowa, and as far east as Huntington, Indiana. In 1965, NH_4NO_3 was the second most popular carrier and urea was third. By 1978, there was greater consumption of urea than of NH_4NO_3 in the U.S. These three carriers are the basis for making N solutions, which have increased steadily since 1965. These trends are shown in Figure 11.1. Today, 95% or more of the N in fertilizers is produced by direct synthesis of NH_3, anhydrous ammonia.

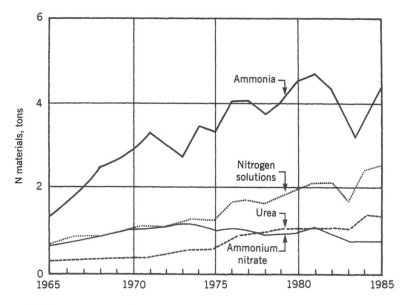

Figure 11.1 Consumption, millions of short tons, of selected nitrogen materials for direct application in the U.S. (Courtesy National Fertilizer Development Center, TVA.)

11.1.2 Anhydrous Ammonia

Modern NH_3 synthesis had its beginning when Haber and Bosch constructed the first commercial plant in Oppau, Germany, in 1913. The process reacts H and N directly, using high temperature and pressure in the presence of an iron catalyst. The reaction is

$$3H_2 + N_2 = 2NH_3 \tag{11.1}$$

Today, over 80% of the plants use the steam-reforming method. Natural gas, CH_4, is the feedstock for H, and the N is obtained from the atmosphere. A flowchart of the major steps is given in Figure 11.2. In the first step steam, CH_4, and air are reacted or reformed to produce H_2, N_2, and CO. These gases are cooled

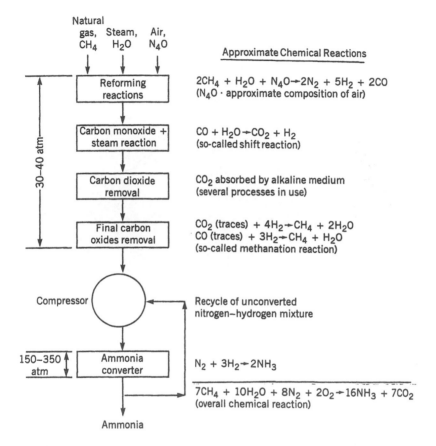

Figure 11.2 Flowchart of ammonia synthesis. (Courtesy National Fertilizer Development Center, TVA.)

and reacted with additional water to produce more H_2 by the shift reaction. The CO_2 produced in the shift reaction is removed by absorption in an alkaline solution. Any carbon-oxygen compounds present must be removed because they poison the catalyst, and this is accomplished in the methanation step. The gases leaving methanation contain about 74% H_2, 0.8% CH_4, and 0.3% argon on a dry basis. Ammonia synthesis occurs when the N_2 and H_2 gases are passed over an iron catalyst, promoted with potassium and alumina, at high temperature and pressure.

At normal pressure, NH_3 is a gas and is transported and handled locally as a liquid under pressure. Specialized equipment is required to handle it. It is injected into the soil to prevent serious loss through volatilization (Figure 11.3).

The gaseous NH_3 protonates to form NH_4^+ in the soil and becomes exchangeable NH_4^+, which is stable. Gaseous ammonia can be lost, depending on three factors. Studies have shown that volatilization losses at pH 7 are low, but increase 10 times when the pH is increased to 8, and 50 times when the pH is increased to 9. When NH_3 is injected into a soil at field capacity, little NH_3 is lis lost at

Figure 11.3 Anydrous ammonia injection into soil before planting corn.

any depth of injection. But when the soil is either drier or wetter than field capacity, considerable NH_3 is lost if the injection depth is less than three inches. The interaction between moisture content of soil and depth of injection on NH_3 loss are shown in Figure 11.4.

11.1.3 Urea

Urea is the second most-popular N carrier and the most popular dry carrier. It requires no specialized equipment for storage and handling and is preferred over NH_3 in situations where the fertilizer is applied by broadcast on lawns or field crops. But it should be noted that broadcast urea may lose considerable N if it is not dissolved by rain (or irrigation) within a few hours after application.

The CO_2 from ammonia synthesis is used to produce urea:

$$2NH_3 + CO_2 = NH_2COONH_4 \ (\text{carbamate}) \tag{11.2}$$

$$NH_2COONH_4 = NH_2CONH_2 \ (\text{urea}) + H_2O \tag{11.3}$$

Urea is more expensive than NH_3, because there are additional manufacturing steps and greater transportation costs, because urea has a lower N content (46%).

A method for making urea prills or granules uses the falling-curtain technique developed at Tennessee Valley Autority (TVA). Small seed urea granules are placed on longitudinal shelves in a rotating drum and are sprayed with a concentrated urea solution as the drum rotates and the granules fall from one shelf to another. Granules increase in size with time. The technique is efficient in producing a wide variety of particle sizes. The dried prills or granules are coated

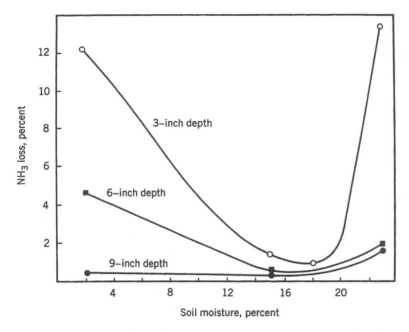

Figure 11.4 Losses of ammonia from Putnam silt loam as influenced by soil moisture and depth of application of 112 k of nitrogen per hectare (100 pounds per acre) in rows 100 cm apart. (From Stanley and Smith, 1956. Used by permission of the Soil Science Society of America.)

with a conditioner to prevent the hygroscopic urea from taking up water from the air and caking (to a solid mass).

When urea is applied to the soil, the hydrolysis of urea is catalyzed by urease:

$$CO(NH_2)_2 + H_2O + urease = 2NH_3 + CO_2 \tag{11.4}$$

Urease is produced by soil microorganisms, but is ubiquitous because the enzyme persists in soils after the enzyme-producing organisms die.

In effect, the application of urea is similar to the application of NH_3, and the N in urea is similarly subjected to volatilization loss. Surface applications of urea, in the absence of movement into the soil by water, commonly result in losses of 20 to 30% of the N. On the other hand, urea is widely used to fertilize rice in flooded fields where anaerobic conditions inhibit nitrification and the subsequent loss of N by denitrification. Superlarge urea granules with a diameter of about 2 cm have been developed for hand application deep in the mud of rice paddies in Asia.

Biuret, NH_2–CO–NH–CO–NH_2, is formed during urea synthesis and in the processing of solutions following urea synthesis. The amount of biuret formed depends on temperature and other conditions, and is commonly 0.8 to 2.0% of the finished product. For most plants, this level of biuret is not toxic when urea is placed in the soil, for the biuret is readily decomposed and its N becomes available. If urea is placed with or close to seeds, they may be damaged, depending on the biuret content of the urea. Biuret is also toxic to citrus plants and some other crops if urea is used as a foliar spray. Citrus is able to tolerate up to 0.25% biuret in urea used for foliar application. Corn and soybeans can tolerate solutions made with urea containing 1.5% biuret.

11.1.4 Ammonium Nitrate

Ammonium nitrate is produced by oxidizing NH_3 to HNO_3 and neutralizing the nitric acid with more NH_3. Both urea and NH_4NO_3 can be made by the falling-curtain technique or by spraying concentrated solutions from the top of a tower and collecting the congealed particles at the bottom. Both processes require that the prills be dried and coated to prevent the adsorption of water from the air and caking. Ammonium nitrate is 33% N with half of the N in nitrate form and the other half in ammonium form. If NH_4NO_3 is applied on the surface, less NH_3 is lost to volatilization, as compared to urea and the substance is well suited for top dressing lawns and other surface applications. After ammonium nitrate moves into the soil, however, the loss of N by NO_3^- leaching and denitrification is more likely.

Ammonium nitrate mixed with 6% fuel oil is used for blasting in mines. Under proper conditions, it is explosive but poses no threat in its normal use as a fertilizer. Thin layers of NH_4NO_3 by themselves will not burn unless continuous heat is applied. Large piles in bulk storage or as bags, however, may become sufficiently heated to ignite under conditions of confinement and gas pressure; they may, if ignited, transmit fire rapidly through the mass and possibly cause an explosion. Even so, there are international trade restrictions on NH_4NO_3, which have helped make urea the most popular dry N fertilizer worldwide. In some countries, the use of ammonia nitrate is prohibited, and it is commonly mixed with lime to produce a material that contains 75% NH_4NO_3 and has a N content of 26%. It is called calcium carbonate-nitrate, CAN.

11.1.5 Nitrogen Solutions

Anhydrous ammonia, urea, and ammonia nitrate are soluble in water and are used to make fertilizer solutions. Solutions using NH_3 have vapor pressure and are used primarily as a source of N in the manufacture of compound fertilizers. Solutions containing only NH_3 are called *aqua ammonia*. Solutions without NH_3 have no vapor pressure and can be used as dry materials are used, except that they are liquid. It is easy to incorporate herbicides at the time of application and to meter the fertilizer solution into irrigation water.

Nitrogen solutions are designated to indicate their composition. A solution of 28.0 (0-40-30) contains 28.0% N, 0% NH_3, 40% NH_4NO_3, and 30% urea by weight. The material contains 30% water. The concentration of N is limited by the crystallization or salting-out temperature. The composition and properties of three common N solutions are given in Table 11.1.

Table 11.1 Characteristics of Urea-Ammonium Nitrate Solutions

Grade, percent N	28	30	32
Composition by weight, percent			
Ammonium nitrate	40.1	42.2	43.3
Urea	30.0	32.7	35.4
Water	29.9	25.1	20.3
Specific gravity, 60°F	1.28	1.30	1.32
Salting-out temperature, °F	1	14	28

Source: Data from Fertilizer Manual, International Fertilizer Development Center, TVA, 1979.

11.1.6 Controlled Release Fertilizers

The N fertilizers that have been discussed are water-soluble, and the N may be quickly absorbed by plants or lost from the soil. For the maintenance of turf, this rapid dissipation of the N requires several applications of N fertilizer annually. Thus, there is a need for controlled-release N fertilizers to reduce the cost of application and increase the efficiency of N use, which is seldom greater than 50%. Controlled release is achieved by making insoluble materials or coating soluble materials.

Sulfur-coated urea, SCU, has been under development at TVA for over twenty years. Urea is the least expensive major dry N carrier. Sulfur is also relatively inexpensive, and the final product has good handling characteristics. Urea granules are coated with sulfur and then coated with a sealant. The thickness of the S coating affects the rate of N dissolution. A total coating of 20 to 22% results in the dissolution or release of 20 to 25% of the N in the first week. SCU-10 and SCU-40 release 10 and 40% of their N by dissolution in seven days, respectively. SCU costs about 35 to 40% more than urea per pound of N and is too expensive for general agricultural use. It is has been found to be effective in reducing N loss for rice production under rain-fed conditions where soils are subject to periodic flooding and drying.

Urea-formaldehyde, ureaform, is the reaction product of urea and formaldehyde in aqueous solution. Ureaform commonly contains 38% N, of which 30% is soluble in water at 25°C. It is desirable to have 50 to 70% of the remaining N soluble in boiling water.

Isobutylidene diurea, IBDU, is produced by the reaction of urea with isobutyraldehyde and contains 32% N. Its production is mainly in Japan and Germany.

Other controlled-release N fertilizers are being developed and tested in various countries. All of these materials are more expensive than urea and are used mainly on lawns, golf courses, horticultural gardens, and for the production of some fruits and vegetables. Controlled-release materials are a component of many lawn and garden fertilizers.

11.1.7 Other Nitrogen Fertilizers

Ammonium phosphates are produced by the neutralization of phosphoric acid with ammonia. They are important as both N and P fertilizers and are considered later in the P fertilizers part of this chapter.

Ammonium sulfate was once the leading N fertilizer produced as a byproduct of the coking of coal in the manufacture of steel. It is still produced as a byproduct and in some instances is manufactured by neutralizing H_2SO_4 with NH_3. The principal disadvantage is its low N content, which is 21%. Its advantages include its sulfur content and good physical condition. The fertilizer is strongly acidic, which can be a disadvantage when used on acid soils or an advantage when used on alkaline soils where the acid effect may be desirable for azaleas, and other acid-loving species.

Calcium nitrate, 15.5% N, is produced by reacting HNO_3 and $CaCO_3$. The nitrate is a readily available source of N and used for soil application on winter season vegetables and as a foliar spray on both vegetables and fruits. Potassium nitrate, 13% N, is manufactured by reacting HNO_3 with KCl and is used in the production of fruits and vegetables.

11.1.8 Nitrogen Carrier Comparisons

Nitrification of ammonium in soils tends to result in the uptake of N as nitrate; thus, many short-term experiments conclude that ammonium and nitrate forms are comparable for increasing yields on a per-pound-of-N basis. In flooded soils, ammonium N is superior to nitrate in increasing yields because nitrate loses N through denitrification. On the other hand, ammonia and ammonium forms applied on or at the soil surface are more subject to volatilization loss than nitrate. The major practical difference in N fertilizers is likely in differences in their effect on soil pH.

Nitrification is an important soil-acidifying process; when ammonium fertilizers are used, there is considerable potential to increase soil acidity. The differences in acidity and basicity between ammonium and nitrate fertilizers was discussed in Chapter 5 under "Culturally Produced Soil Acidity."

There is great variation in the partitioning of added fertilizer N as climate, soils, crops, yields, and soil management practices vary. Some estimates of the fate of N fertilizer are 30 to 70% removed in harvested crop, 5 to 10% leaching loss, 10 to 30% gaseous loss, and 10 to 40% incorporated into soil organic matter. As a general rule, 50% of fertilizer N is absorbed by the crop; 25% is lost by denitrification, leaching, and volatilization; and 25% remains in the soil as mineral N or incorporated into new organic matter.

11.1.9 Organic Nitrogen Fertilizers

Manure, crop residues, and many other organic materials have been added to soils to increase the nutrient supply. The use of sludge and other waste products has increased considerably in the last two decades because it has been environmentally correct to recycle nutrients in these waste materials rather than placing them in landfills. The benefits come largely from their N content. Nitrogen in organic form can be stored in the soil; however, the N is available only after mineralization. Historically, nonleguminous crops have been rotated with legumes so that they can take advantage of the N resulting from the decomposition of legume residues. A cover crop that grows in the fall and early spring takes up mineral soil N and converts it into organic matter, thereby reducing leaching and denitrification losses. Plowing under the cover crop allows a gradual release of N by mineralization during the growing season.

The advent of cheap N fertilizer has greatly reduced the importance of organic N fertilizers in the world during the 20th century. Given present concerns about the environment and interest in sustainable agriculture systems, organic fertilizers may be more sought-after in the future.

11.2 PHOSPHORUS FERTILIZERS

The beginning mineral for most, if not all, of P fertilizers is apatite, or rock phosphate. Large phosphate rock deposits exist in Florida, where they are buried under various kinds of overburden, as shown in Figure 11.5. Large deposits also exist in Morocco, Russia, and China. The mineral deposit is exposed generally by strip mining. Then the mineral deposit is removed by large shovels, drag lines or water under pressure. The mineral is seldom in a pure form and must be separated from impurities, including inert material, before it is utilized. Once it is separated from impurities, rock phosphate may be finely ground and marketed directly as a fertilizer source of P. This product generally contains from 11.5 to 17.5% total P. None of the P will be water soluble, but from 5 to 17% of the total P is soluble in citric acid. The term *available* P is applied to P fertilizers and includes the citric acid soluble P. When rock phosphates undergo a manufacturing process to make a more soluble P fertilizer, two alternative approaches exist: first, direct acidification of rock phosphate to produce superphosphates; second, the product is phosphoric acid or elemental P, which becomes a basic material for further manufacture.

11.2.1 Superphosphates

If sulfuric acid is mixed with rock phosphate, the P is rendered soluble by the following reaction:

$$Ca_{10}F_2(PO_4)_6 + 7H_2SO_4 = 3Ca(H_2PO_4)_2 + 7CaSO_4 + 2HF \qquad (11.5)$$

The product is commonly called superphosphate, 0-20-0, and contains soluble monocalcium phosphate and calcium sulfate. The HF is a toxic gas that escapes during manufacturing and usually is (and certainly should be) recovered as a byproduct. Although this product is less common in the U.S. today because of

Figure 11.5 Location and nature of Florida rock phosphate deposits. (Used by permission of the Florida Phosphate Council, Inc.)

the expense of shipping a low-analysis fertilizer, it is an excellent source of P, which carries both Ca and S in addition to P.

Partially acidifying rock phosphate is a practice which merits consideration in developing countries where sophisticated manufacturing may not be possible. Waste sulfuric acid, which is often available, can be used to partially react with rock phosphate, producing a material which contains considerable soluble P. Although it is not as effective as 0-20-0, it is much more useful than direct application of rock phosphate.

Acidification of rock phosphate with phosphoric acid produces a high-grade, water-soluble product known as triple superphosphate, 0-45-0, according to the following reaction:

$$Ca_{10}F_2(PO_4)_6 + 14H_3PO_4 = 10Ca(H_2PO_4)_2 + 2HF \qquad (11.6)$$

Although the manufacturing is similar to 0-20-0, this material does not carry gypsum in the final product. It also requires the production of phosphoric acid for manufacture.

11.2.2 Phosphoric Acid Production

Phosphoric acid is prepared by the wet process acid and electric furnace elemental P methods. Wet process acid is prepared by reacting rock phosphate with sulfuric acid, allowing sufficient time for gypsum to crystallize and filtering the material to obtain the phosphoric acid. Although similar to the reaction in Equation 11.5 illustrating the manufacture of ordinary superphosphate, the differences are significant. As shown in the following equation,

$$Ca_{10}F_2(PO_4)_6 + 10H_2SO_4 + 20H_2O =$$
$$10CaSO_4 \cdot 2H_2O + 6H_3PO_4 + 2HF$$

$$(11.7)$$

additional sulfuric acid and water are added to the rock phosphate. Usually about eight hours are allowed in the digestion mixture, after which time the crystals of gypsum, $CaSO_4 \cdot 2H_2O$, are sufficiently large to be filtered easily. This acid may be directly used for acidification of rock phosphate but requires further purification and concentration before it can be used for the manufacture of polyphosphate fertilizers.

Electric furnace phosphoric acid is prepared by heating rock phosphate to temperatures in excess of 1,400°C with carbon and silica in a reducing atmosphere. The reaction produces elemental P, which may be oxidized to produce P_2O_5, which when hydrated produces phosphoric acid. This will produce a high purity acid and, with proper control, will produce acid of the desired ratio of orthophosphate to polyphosphate for use in liquid fertilizer manufacture. But due to the more expensive manufacturing process, most of the phosphoric acid that goes into fertilizer manufacture is prepared by the wet process method.

11.2.3 Ammonium Phosphates

In addition to manufacturing 0-45-0 which was discussed previously, phosphoric acid is utilized for manufacture of monoammonium phosphate (MAP), diammonium phosphate (DAP) and ammonium polyphosphate (APP). Ammonium orthophosphate can be prepared to yield two products, MAP and DAP, depending on the degree of ammonification. Several methods are used to manufacture the products. Basically, phosphoric acid is treated with ammonia. If the neutralization is carried out to a limited extent, MAP is produced with a typical analysis of 11-48-0 as follows:

$$NH_3 + H_3PO_4 = NH_4H_2PO_4 \tag{11.8}$$

If neutralization is carried out to obtain a higher ratio of ammonia to phosphoric acid, DAP is produced with a typical analysis of 18-48-0 as follows:

$$2NH_3 + H_3PO_4 = (NH_4)_2 HPO_4 \qquad\qquad (11.9)$$

Grades between the two mixtures can be obtained with intermediate ammonification.

With the development of superphosphoric acid came the possibility of producing higher-analysis ammonium phosphates, usually called ammonium polyphosphates, APP. The most commonly used superphosphoric acid contains about 50% orthophosphate with the remainder as higher polymers of phosphoric acid. The major portion of the higher polymers are as pyrophosphates, making up 43% of the total P. Ammonium polyphosphates have formed an important part of the liquid fertilizer industry. A rather simple process of carrying out the neutralization reaction in a pipe reactor allows for the preparation of the liquid fertilizer on-site by neutralization of the superphosphoric acid with ammonia. Typical grades are 11-32-0 and 11-37-0. A typical grade of APP, when the fertilizer is in the dry form, is 15-62-0.

11.3 USE OF PHOSPHORUS FERTILIZERS

11.3.1 Use of Rock Phosphate

Rock phosphate may be directly applied to soils, usually by broadcasting and incorporating into the soil with tillage. Years of research work several decades ago has shown that this material is not effective when utilized in a management scheme where soils are limed to maintain soil pH above 6.0. As shown earlier in Chapter 7 and in this chapter, the solubility of rock is so limited that it is of little use under slightly acid to neutral conditions.

Rock phosphate may be of considerable value, however, if applied to highly weathered acid soils which are most often very low in available P. Rock phosphate added to these soils may be readily soluble, since the pH is low and many times the Ca concentration (activity) is also low. The importance of Ca activity is shown in Figure 11.6. It is also known that Ca activity may not be closely related to pH (Table 11.2). For this reason, results from use of rock phosphate, even on very acid soils, is sometimes unpredictable. But even though sometimes unpredictable in reaction, rock phosphate is the source most-often available in many lesser developed countries. In these countries, the price of importing soluble, manufactured P fertilizers will preclude their use. When rock phosphates are locally available, they may be used to advantage. But it must be emphasized that the rock phosphates will be more soluble on acid soils with low Ca contents. If the management program for these soils includes some application of lime, it should be made after the application of rock phosphate has been made to the soil and has had sufficient time to react, a period of more than six months. Under these conditions rock phosphate may be a very satisfactory fertilizer.

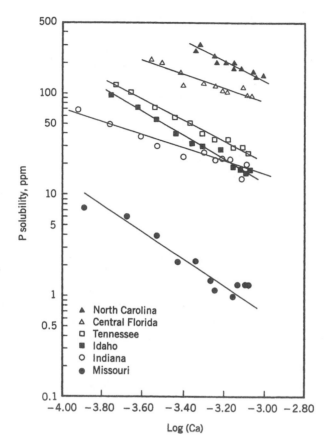

Figure 11.6 The effect of calcium ion activity (mol/L) on the solubility (ppm P) of six rock phosphates. (From Wilson and Ellis, 1984. Used by permision of the Williams & Wilkins Company.)

Table 11.2 Calcium Ion Activities in Soils

Soil	pH	Ca²⁺ activity mg/L	m/L
Warsaw loam	5.38	36	9.1×10^{-4}
Munising sandy loam	4.86	35	8.8×10^{-4}
Bradson	4.22	70	1.7×10^{-4}
Tatum	4.94	1.2	3.0×10^{-5}

Source: Wilson and Ellis, 1984.

11.3.2 Use of Superphosphates

Superphosphates are very soluble materials and hence satisfactory for any application which requires a soluble P source. Generally they are excellent sources when coupled with a management program that utilizes liming to maintain soil pH above 6.0 and including calcareous soils. When utilized on acid soil, considerable rapid adsorption and/or precipitation of P will occur when the soluble P reacts with Fe and Al that is dissolved or in the amorphous form in soils. Often these materials will be band placed near the seed at planting time to slow down the process of fixation. In this placement technique, the band of fertilizer will become very acid as the fertilizer adsorbs water from the surrounding soil and dissolves. The pH in the band may become much lower than 3.0. Although root growth cannot penetrate such an acid band, the P diffusing from this band will be in a high concentration and readily available for plant uptake. A residue of dicalcium phosphate, $CaHPO_4$, has been shown to remain in the band.

When these soluble materials are applied to a calcareous soil, particularly if incorporated, they will react to form first dicalcium dihydrate, $CaHPO_4 \cdot 2H_2O$, then the anhydrous form, $CaHPO_4$, will appear, followed in a few weeks by a less soluble octacalcium phosphate, $Ca_4H(PO_4)_3$. This form is stable for months or years, but evidence suggests that β-tricalcium phosphate, β-$Ca_3(PO_4)_2$, will form, given sufficient reaction time. Both forms of dicalcium phosphate and octacalcium phosphate are sufficiently soluble and dissociate with enough rapidity to furnish P for plant uptake; consequently, this conversion to the less-soluble P compounds is not a loss to soil fertility. Further conversion to β-tricalcium phosphate and apatite, if the latter conversion indeed occurs, leaves P in forms that are not sufficiently soluble for optimum P availability in calcareous soils.

11.4 FATE OF APPLIED FERTILIZER PHOSPHORUS

The desire and goal of every farmer is to have a high percentage of fertilizer applied utilized by the current crop. But with P this is far from reality. The uptake of P may be from 5 to 30% of that applied during the current year. The remainder converts to insoluble compounds, adsorbed P, or organic P. These compounds may be available for crops grown in later years.

11.4.1 Initial Reactions of Soluble Fertilizer Phosphorus

Most fertilizers are hygroscopic, so they attract water once added to soils. Water may move to the fertilizer granule either by capillary adjustment or by vapor movement, but the result is the formation of a saturated solution of fertilizer material in water. The initial reactions will then be dependent upon the characteristics of the fertilizer. The pH of the soil surrounding the fertilizer particle reflects the pH of a saturated solution of the material. The wide variance that may be obtained is shown in Table 11.3.

When monocalcium phosphate, 0-45-0, is added to a soil, it forms a very acid solution. If the individual fertilizer particles have been broadcast and incorporated into the soil, this acidity is soon neutralized by the soil. But if the fertilizer material has been placed in a band, the soil surrounding the band will become very acid and may have a pH as low as 2.2. Precipitation occurs within the fertilizer band giving a residue of dicalcium phosphate, $CaHPO_4$. The saturated

Table 11.3 Content and pH of Saturated
Solutions of Fertilizer

Compound	Saturated solution	
	pH	P, mol/L
$Ca(H_2PO_4)_2 \cdot H_2O$	1.0	4.5
$(NH_4)_2HPO_4$	8.0	3.8
K_2HPO_4	10.1	6.1

Source: Sample, Soper, and Racz, 1980.

solution of P will move slowly into the soil, where P may be absorbed by plants or utilized by microorganisms if the extreme pH and high ionic strengths are not detrimental to the organisms. The P may also be adsorbed by soil particles or converted to less-soluble compounds.

The exact nature of these reactions are controlled by the soil properties and by the placement of the fertilizer. If the fertilizer is broadcast and mixed with a large volume of soil, it will dissolve quickly and the P will reach an equilibrium that is governed by the properties of the soil. For calcareous soil, this will include adsorption by $CaCO_3$ or Fe oxides/hydroxides and precipitation as dicalcium phosphate, octacalcium phosphate, and/or tricalcium phosphate. In acid soils, adsorption by Fe and Al oxides/hydroxides and clay minerals is thought to be the most important reaction. Precipitation as Fe and Al phosphate compounds will also occur.

Banding the fertilizer material will delay the time for the fertilizer to reach equilibrium with the soil. This has been particularly advantageous when applying P to soil that is very low in available P. Less P will be required to reach a satisfactory yield with band placed as opposed to broadcast and mixed. Early season growth effects may also be obtained by band placement of fertilizer, even in soils that are high in available P. The ultimate fate of fertilizer that is band placed is to reach equilibrium with the soil. When plowing was a standard tillage practice, even the banded fertilizer was well mixed with the soil once a year. But with zero-tillage methods, this may no longer be the case. Residual effects of band-placed fertilizer may persist for relatively long periods of time under reduced tillage systems.

11.4.2 Residual Soil Phosphorus

A number of important questions arise concerning residual soil P:

1. What is the fate of P that is converted to a residual form in soils?
2. Do common soil test methods accurately assess residual P?
3. Should additional fertilizer P be added for crop growth even if residual P is high?

Much of the residual P remains in a labile form. It may be as adsorbed P or precipitated P. A portion of this P is measured by most soil-testing procedures. But some soils contain more P in the residual form than can be removed by repeated cropping or than is measured by soil test. For most agronomic crops, when a soil contains sufficient residual P so that the test is high, little benefit is gained from additional fertilizer P, even in the form of band placed at planting time. Certain crops are an exception to this rule: potatoes, for example, may give a yield response to added P fertilizer even when the soil test levels are very high.

11.4.3 Effect of No-Till on Soil Phosphorus

Two conditions occur with no-till systems that have an important bearing upon soil P: first, P applications must be made to the soil surface or applied in a band close to the surface; second, organic matter is deposited on the surface of the soil rather than being incorporated into the plow layer.

Surface and band applications of P fertilizer without cultivation lead to non-uniform distribution of P in the soils. This is of little concern if the soil initially has a high level of available P. Since most management systems have just recently converted to no-till systems, many soils presently have high levels of P to a considerable depth in the profile. It is likely that it will take many years of no-till management before P is removed from the upper layers of the soil and deposited on the surface. But the ultimate end would be a distribution of P that is less than desirable for crop growth, in that root growth is likely to be much better in the layers which contain the high levels of P.

Accumulation of organic P in the surface few centimeters of soils under no-till systems has the effect of increasing soluble and labile P in the layer most exposed to runoff, which will increase the P in the runoff water. This may be relatively unimportant in that the same tillage systems are very effective in reducing surface runoff so that the total P which leaves a field will be greatly reduced by conservation and no-till systems.

11.5 POTASSIUM FERTILIZERS

Wood ashes that contained K_2CO_3 were an early source of K for industrial and agricultural use. The first commercial-scale production of K fertilizers began in Germany in 1861, only about fifteen years after Liebig discovered that K was an essential element for plants. Until then, salts had been mined in Germany for NaCl and the accompanying K salts were discarded (Salzburg means "salt city"). Because K fertilizers were needed, production developed rapidly in Germany and dominated world trade until after World War I.

11.5.1 Potassium Ore Deposits

Most of the K ore is now obtained by mining evaporite deposits that were created by the evaporation of ancient seas under arid conditions. These deposits exist under overburdens of varying thickness. More than half of the world's known reserves occur in Canada, Ukraine, Byelorussia, and the Urals in Russia. Other major deposits exist in the U.S., Europe, the Middle East, Thailand, Zaire (the Congo), and South America.

The major mining operations in the U.S. are in southern New Mexico near Carlsbad. Here, evaporite deposits occur in the Permian basin. Many ore layers are present, with the most productive zone being 1.5 to 3 meters thick, at 250 to 610 meters below the surface. The ore is shaft mined. Other important operations include solution mining of evaporite deposits in Utah and recovery of brine deposits in Utah and California. Canada has the largest K reserves in extensive deposits in Saskatchewan. The evaporite ore is 660 to 3,050 meters below the surface. In places the ore is too deep for shaft mining, so solution mining is then used. These deposits extend southward in North Dakota where they are deeper and not mined. About 95% of all the K mined is used for fertilizer manufacture.

Many different salts are found in evaporite deposits. At Carlsbad, the ore is mainly composed of sylvinite, a physical mixture of KCl and NaCl crystals. Present to a lesser degree is langbeinite, composed of sulfates of K and Mg. Nearly all of K used for fertilizers comes from the processing of sylvinite and basically requires a separation of KCl and NaCl. Over 95% of K fertilizer is KCl and has a K content of at least 49.8%. Three methods used for production of KCl using sylvinite ore are flotation, heavy media separation, and fractional crystallization.

11.5.2 Fractional Crystallization Method

Fractional crystallization is a historic method developed in Germany. The separation of KCl from NaCl is based on the different solubilities and temperature characteristics of the two salts. The solubility of NaCl remains quite constant with temperature, whereas the solubility of KCl is greatly increased by an increase in temperature. Saturated solutions of both salts at ambient temperature are heated to boiling. Then, enough crushed ore is added to saturate the solution in respect to KCl. The dissolution of KCl depletes the ore of KCl, but the NaCl content of the ore is little affected. The hot brine is cooled to ambient temperature to crystallize out the KCl that dissolved from the ore. The recrystallized KCl is separated from the brine, dried, and sent to storage.

11.5.3 Froth Flotation Method

The froth flotation method is the most popular and economical. Other advantages include ease of control and flexibility. Figure 11.7 is a diagram of the flotation process.

The ore is crushed fine enough so that relatively pure grains of NaCl and KCl are produced. These are added to a saturated brine of KCl and NaCl, which produces a pulp containing 50 to 75% solids. Desliming separates foreign substances, such as clays and other insoluble materials. The ore is then treated with hydrogenated tallow amine, a hydrochloride soap, which preferentially adheres to the KCl grains. Air is bubbled through the pulp in shallow, agitated tanks or flotation cells. The KCl particles float to the top and are continuously skimmed off. The NaCl and other salts from the ore remain on the bottom of the cells and

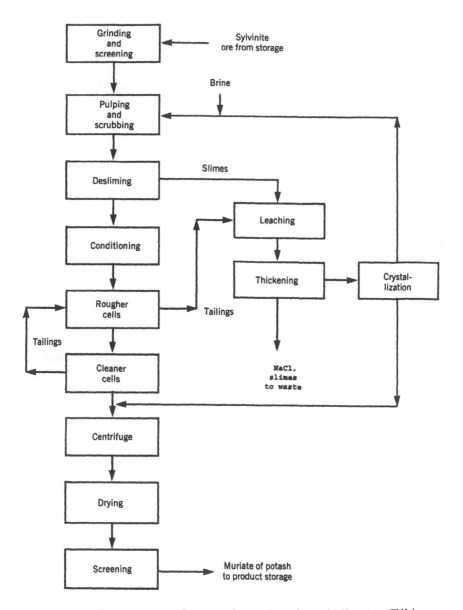

Figure 11.7 The flotation process for recovering muriate of potash. (Courtesy TVA.)

are continuously removed. The KCl particles are filtered from the brine, dried, and sent to storage.

11.5.4 Heavy-Media Separation

The heavy-media separation method uses large or coarse crystal grains, which are found in the Saskatchewan ore deposits. Less grinding is required, and the

KCl is separated from NaCl, based on difference in specific gravity. The ore is coarse-crushed and particles in the 1.6- to 6.5-mm size range are removed and used, while the smaller particles are processed by the froth flotation method.

The heavy-media process uses a saturated brine to which magnetite, Fe_3O_4, with a specific gravity of 5.17, is added. Enough magnetite is added to produce a specific gravity of the saturated brine between 1.98 and 2.13, the specific gravity of KCl and NaCl, respectively. Sized particles in the range of 1.6 to 6.5 mm in diameter are mixed with the specific-gravity-adjusted brine. Centrifugal separation methods are used to remove the KCl, which is debrined, dried, crushed, and sent to storage.

11.5.5 Sizing and Granulation

Elaborate screening and granulation facilities are used to produce particles in sizes appropriate for various uses. Bulk blenders require a particle size that matches the other materials used in the blend; the particles should have diameters in the range 0.84 to 3.36 mm. Suspension fertilizer manufacturers need particles of KCl in the range 0.11 to 0.42 mm. Drum granulation to produce complete or mixed fertilizers uses particles in the 0.2- to 1.2-mm size range. Some particle size needs are met by passing KCl from the primary refinery operation through high-pressure compactor rollers. The dense KCl flakes produced are crushed and screened to provide the desired particle size (granulation). Undersized particles are recycled.

11.5.6 Other Processes and Materials

Most commercial K fertilizer plants produce K_2SO_4 by reacting KCl with H_2SO_4. Langbeinite, $K_2SO_4 \cdot 2MgSO_4$, is used as a source of K, Mg, and S. Reacting langbeinite with KCl produces K_2SO_4 and $MgCl_2$. Reacting KCl with HNO_3 produces KNO_3. Small amounts of K are produced from natural brines at Searles Lake in California, Great Salt Lake in Utah, and the Dead Sea in Israel. At Great Salt Lake and the Dead Sea, brine is taken from the area where water has been retained the longest and brine concentration is the greatest. The brine is pumped into solar evaporation ponds, where water evaporates and salt crystallizes.

REFERENCES

Barber, S. A., R. D. Munson and W. B. Dancy. 1985. Production, Marketing and Use of Potassium Fertilizers. *Fertilizer Technology and Use*. Soil Sci. Soc. Am., Madison, WI.

Engelstad, O.P., Ed. 1985. *Fertilizer Technology and Use*. 3rd ed. Soil Sci. Soc. Am., Madison, WI.

Follett, R. H., L. S. Murphy and R. L. Donahue. 1981. *Fertilizers and Soil Amendments*. Prentice-Hall, Englewood Cliffs, NJ.

Hoffmeister, G. 1979. *Physical Properties of Fertilizers and Methods for Measuring Them.* Bull. Y-147. TVA, Muscle Shoals, AL.

International Fertilizer Development Center. 1979. *Fertilizer Manual.* TVA, Muscle Shoals, AL.

Jackson, J. E. and G. W. Burton. 1958. An Evaluation of Granite Meal as a Source of Potassium for Coastal Bermudagrass. *Agronomy J.* 50:307-309.

Jones, U. S. 1982. *Soil Fertility and Fertilizers.* 2nd. ed. Prentice-Hall, Reston, VA.

National Fertilizer Development Center. 1983. *Fertilizer Trends 1982.* TVA, Muscle Shoals, AL.

Pierre, W. H. 1933. Determination of Equivalent Acidity and Basicity of Fertilizers. *Industrial and Eng. Chem.* 5:229-234.

Rader, L. F., L. M. White and C. W. Whittaker. 1943. The Salt Index: A Measure of the Effect of Fertilizers on the Concentration of the Soil Solution. *Soil Sci.* 55:201-208.

Sample, E. C., R. J. Soper, and G. J. Racz. 1980. Reactions of Phosphate Fertilizers in Soils. In *The Role of Phosphorus in Agriculture.* F. E. Khasawneh, E. C. Sample, and E. J. Kamproth, Eds. Am. Soc. Agron., Madison, WI.

Stanley, F. A. and G. E. Smith. 1956. Effect of Soil Moisture and Depth of Application on Retention of Anhydrous Ammonia. *Soil Sci. Soc. Am. Proc.* 20:557-561.

Tisdale, S. L., W. L. Nelson and J. D. Beaton. 1985. *Soil Fertility and Fertilizers.* 4th ed. Macmillan, New York.

Whittaker, C. W. 1948. *Ammonium Nitrate for Crop Production.* Cir. 171, USDA, Washington, D.C.

Wilson, M. A. and B. G. Ellis. 1984. Influence of Calciums Solution Activity and Surface Area on the Solubility of Selected Rock Phosphates. *Soil Sci.* 138:354-359.

Young. R. D. 1983. *Mixed Fertilizers.* Mimeo. TVA, Muscle Shoals, AL.

CHAPTER 12

Mixed Fertilizers

During the early years of fertilizer manufacturing, dry single-carrier fertilizer materials were mixed together. These fertilizers were dusty, and they would segregate during transport and application. The addition of anhydrous ammonia to dry mixes and their subsequent granulation constituted a major advance. The granular fertilizers are dust-free, and each granule has similar composition and particle size. These free-flowing fertilizers were needed for the new fertilizer application equipment that was being developed. More recently, fluid fertilizers have become very popular.

In 1982, it was estimated that 21 million tons of mixed fertilizers were produced in the U.S. and that this represented 44% of the total fertilizer production. Dry granular homogeneous mixtures were the second most-popular, followed by fluid fertilizers. This chapter examines the composition, manufacture, and properties of the major mixed fertilizers used today.

12.1 MAJOR FERTILIZER SYSTEMS

There are two overlapping organizations in the production and marketing of fertilizers in the U.S. As discussed in earlier chapters, the major N material, NH_3, is produced near natural gas well-heads or pipelines. The major P source materials for fertilizers — superphosphates, ammonium phosphates and phosphoric acid — are produced near the rock phosphate mines in Florida, North Carolina, and some western states. Most of the KCl was produced near Carlsbad, New Mexico, until the 1960s when mines were opened in Saskatchewan. Now, 70% or more of the K in fertilizers in the U.S. comes from the Canadian mines. These materials or carriers are frequently applied directly to soils, but they are called intermediates to indicate their role in the manufacture of mixed fertilizers. These fairly concentrated intermediates are transported to regional and local plants near markets, and there they are combined into mixed fertilizers. The production pattern for finished mixed (NPK) fertilizers in the U.S. has evolved into three main systems:

granulation, bulk-blending, and fluid fertilizers. The relative importance of these three systems in shown in Table 12.1.

Table 12.1 Consumption of Granulated, Bulk-Blended, and Fluid Fertilizers

Type	Total fertilizers, percent	Total mixed fertilizers, percent
Granular NPK	16.6	28.6
Bulk-blended	31.5	54.3
Fluid	9.9	17.1

Source: Unpublished TVA data for the U.S., 1981.

12.2 GRANULAR FERTILIZERS

The process of granulating dry fertilizers was developed to provide a more uniform product for use in labor-saving mechanical applicators and to reduce costs of manufacturing and transport. At first, dry-mixed fertilizers are moistened with water and subjected to mechanical action in a mixer to form more or less uniform-sized particles. After the batch is granulated in the mixer, the particles are dried, screened, and cooled. Oversized and undersized particles are recycled. A coating material is added to minimize lumping and caking during storage. This method of granulation is suitable for small plants and is still important in developing countries.

In time, the continuous drum or tube granulator, as opposed to the batch mixer, became popular. The granulator is a rotating drum about 1 to 2.5 m in diameter and 3 to 10 m long. Incoming materials are screened to remove lumps, which are crushed. The materials are weighed and continuously fed into the granulator at a controlled rate. Steam is discharged under the bed of material and water is sprayed on top. Granulation is controlled by the amount of water added. Discharged particles are dried, cooled, and screened with the oversized and undersized granules recycled. A flow diagram of a typical granulation plant is shown in Figure 12.1, and some typical formulations for grades are given in Table 12.2.

In efforts to increase the N content of granular fertilizers, an ammoniator was added to the process and, more recently, phosphoric and sulfuric acids were added. As a result, many chemical reactions occur in the granulator and both dry materials and slurries are granulated.

12.2.1 Melt Granulation

The melt granulation process developed in the 1960s was first used to produce granular ammonium phosphates. Wet-process superphosphoric acid and ammonia are reacted in a confined pipe in which heat is released, evaporating some of the water that is present and producing a melt. The melt is the liquid phase of the solid fertilizer. The reactor developed by TVA is called the pipe-cross reactor

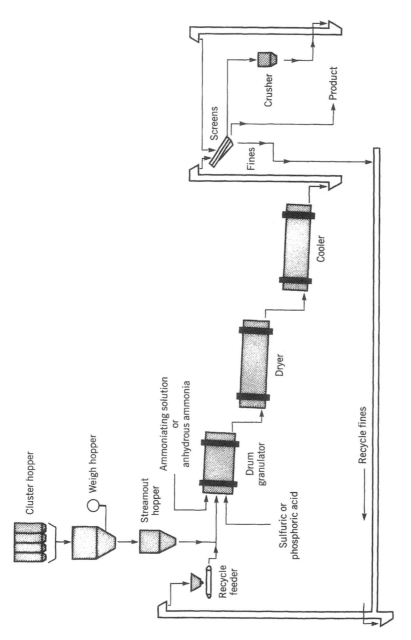

Figure 12.1 A typical layout for granulation plants in the U.S. (Courtesy TVA.)

Table 12.2 Typical Formulations for Some Popular
Grades of Granular Fertilizers

Materials used	12-12-12	6-24-24	10-20-10
Ammonia	—	60	—
44.8% N solution	408	96	457
Ammonium sulfate	294	—	—
Diammonium phospate	—	180	—
Superphosphate, 20%	400	—	420
Superphosphate, 46%	346	864	709
Potassium chloride	400	800	334
Sulfuric acid	140	—	76
Steam	—	150	—

Source: Data from Young, 1983.

(PCR). Phosphoric and sulfuric acid are reacted simultaneously with NH_3 with the release of much heat and formation of a melt. The melt is sprayed into the granulator drum, and with the addition of KCl, a complete mixed fertilizer can be produced (Figure 12.2). Some advantages of the PCR and melt granulation are:

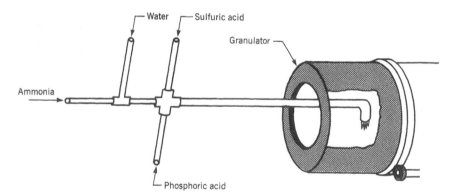

Figure 12.2 Diagram of a pipe-cross reactor used to make melts for the manufacture of granular fertilizer.

1. Reduced costs for drying since granules produced from the melt of the PCR have only 1 to 4% water compared to 10 to 15% for regular granulation.
2. Easier compliance with environmental regulations because fewer ammonium chloride fumes are produced.
3. Easier and more precise control of ingredients.
4. Lower plant construction costs.

The progressive development of better manufacturing methods has resulted in a progressive increase in the nutrient concentration of mixed fertilizers from 15% ($N + P_2O_5 + K_2O$) in 1910 to 44% by 1981. These methods have also achieved a lower cost per pound of plant nutrient.

12.2.2 Caking and Dustiness

Between manufacture and application to the soil, dry fertilizers must be stored in bulk or in bags. It is essential that the fertilizers maintain their free-flowing character and not cake or form lumps in storage. Fertilizer cakes when crystal bonds grow between particles. These crystals may develop during storage as a result of continuing chemical reactions, or thermal effects that produce small crystals from minute amounts of solutions in the fertilizer. Intergrowth of crystals between particles results in cementation or caking. In some cases, cohesive forces between particles have produced caking. Large granules reduce the number of contact points between particles; granular materials have less of a tendency to cake than do pulverized materials. Soft granules under pressure in storage piles may deform, which causes relatively large areas of contact between granules and promotes caking.

A frequent misconception holds that moisture absorbed during storage causes caking, but excessive moisture left in the product during manufacture is the true agent. The N materials have the greatest tendency to cause caking if they are not properly dried. In general, they should contain 0.5% or less of free water (non-hydrate water).

Solid conditioning or anticaking dusts in common use include diatomaceous earth (kieselguhr), kaolin clay, talc, and chalk. Usually 1 to 4% of these powdered conditioners is applied. Clays have good adherence to particles, but diatomaceous earths have greater absorbency. Adherence of conditioners can be improved by spraying the fertilizer with a small amount of oil either before or after the conditioner is applied.

Dustiness is a very undesirable property, especially with our increasing concern about the healthfulness of the work environment. Granulation does much to reduce the dustiness of dry fertilizers, but dustiness may still exist. The dustiness of granular fertilizers is caused by poor sizing, which leaves too many fine, soft granules, which break apart; the poor adherence of conditioners; and formation of surface crystals, which may abrade to form dust. The use of oils just mentioned above is the major way in which dustiness is currently controlled.

12.3 BULK-BLENDED FERTILIZERS

Granulation is used to produce single-nutrient carriers, as well as complete NPK fertilizers. Bulk-blending is the physical mixing of two or more granular materials. The practice was first started in Illinois in 1947. Blended fertilizers can be bagged or distributed in bulk. Only 23% of the blenders in the U.S. have bagging facilities.

A major advantage of bulk-blending is the ease with which many different grades and ratios can be formulated. Prescription mixes guided by soil analysis are popular. The major intermediates used are granular urea, ammonium nitrate, diammonium phosphate, triple superphosphate and potassium chloride. These materials are shipped to a plant located in the area where the fertilizer will be used. The materials are mixed and commonly loaded directly into a truck and immediately spread in the field (Figure 12.3). The entire cycle of weighing, mixing and discharging may be automated. Low cost is a major reason for the popularity of bulk-blending, which now accounts for 54% of all mixed fertilizers in the U.S.

12.3.1 Chemical Incompatibility of Intermediates

Chemical incompatibility of intermediates may cause the mixture to heat up, water to form, gas to evolve, and caking. Only a few materials are troublesome (Figure 12.4). Mixing of urea with ammonium nitrate must be avoided, for it causes a high degree of wetting and is related to the critical relative humidity (CRH). The CRH is the relative humidity above which the material spontaneously absorbs moisture from the air. Relatively pure urea does not absorb water at 70% relative humidity, but it takes up water continuously at 75% relative humidity at 30°C (86°F). This CRH is typical for most fertilizer materials; however, mixtures of salts usually have a lower CRH than either constituent. When urea and NH_4NO_3 with CRHs of 72.5 and 59.4%, respectively, are mixed together, the CRH of the mixture is only 18.1%. Thus, either alone may store well, but when mixed together, they are very hygroscopic and take up moisture from the air at low relative humidity.

Urea reacts with monocalcium phosphate monohydrate in superphosphate to release water of hydration and produce severe stickiness:

$$Ca(H_2PO_4)_2 \cdot H_2O + 4CO(NH_2)_2 = Ca(H_2PO_4)_2 \cdot 4CO(NH_2)_2 + H_2O \quad (11.10)$$

Mixing diammonium phosphate with superphosphates may cause caking during long-term storage.

12.3.2 Segregation

Bulk handling may induce particles of a fertilizer to segregate, whereupon its composition becomes nonuniform throughout. Segregation of particles is undesirable because the consequent nonuniformity can make it impossible to obtain proper samples to meet analytical guarantees. Agronomic responses may be affected and are a particular problem if a micronutrient intermediate is separated from the bulk of the fertilizer.

Segregation occurs because individual particles in a fertilizer respond differently to mechanical disturbances during handling. Particles of similar physical properties tend to congregate, thus destroying homogeneity. Numerous tests have shown that the physical property that is most important in producing segregation is particle size. Neither the density or shape of particles within the ranges found in fertilizers affect segregation to an important degree.

During the flow of a bulk-blended fertilizer into a storage bin or truck, particles of the various intermediates may segregate. The finer particles tend to

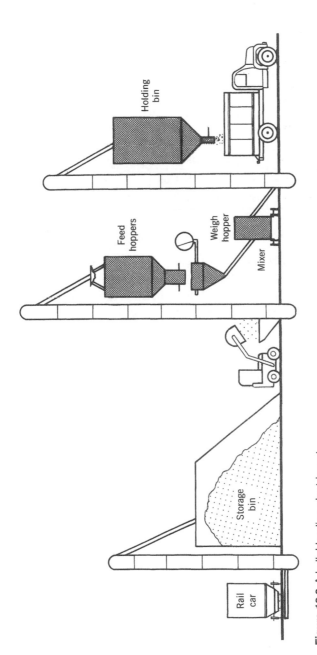

Figure 12.3 A bulk-blending plant layout.

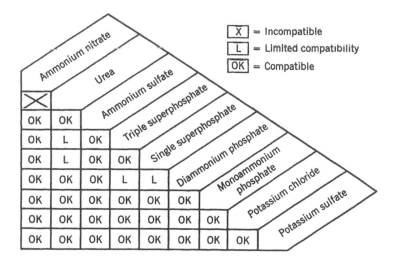

Figure 12.4 Chemical compatibility of blend materials. (Courtesy TVA.)

remain where they fall, while the larger particles roll down the pile toward the perimeter. This process is called coning. Most of the bulk-blended fertilizer is loaded directly into applicators with large flotation tires equipped with a spinner (fan-type) spreader and immediately spread onto fields.

Vibration in the plant after mixing and during transport and spreading may cause particles to segregate. Smaller particles sift downward into the void spaces between larger particles and collect at the bottom. Granules are segregated as they are propelled through the air by fan-type spreaders. Large particles travel farther than the small particles, causing uneven application of nutrients. Thus, particle size is an important consideration in granulation (Figure 12.5). It is recommended that 100% of the materials be within the mesh range of –6 to +16, with at least 25% and not more than 45% of +8-mesh size.

Particle hardness becomes a factor during application because soft granules break up into smaller particles. Prilled urea particles tend to be porous and have the lowest crushing strength of most common materials used for bulk-blending. New granulation methods for urea produce harder granules, compared to prills, and have significantly decreased caking of urea. Particles that remain intact when pressed by the forefinger against a hard surface are considered to have satisfactory hardness.

12.4 FLUID FERTILIZERS

Fluid fertilizers are of two types: solutions and suspensions. The important advantages of producing and marketing fluids are simplicity of the mixing facilities and ease of handling. The fluids themselves are dustless and their other advantages include homogeneity, uniformity of application, the ease with which micronutrients or pesticides are added, the ease with which fluids are applied

DAP+"GRANULAR" KCL   **DAP+"COARSE" KCL**

Figure 12.5 The poor granule-size match on the right caused diammonium phosphate to segregrate from the larger, coarse potassium chloride particles. (Photographs courtesy International Fertilizer Development Center, TVA.)

with irrigation water, and their use as foliar sprays. Disadvantages include lower-analysis products, increased shipping costs, and the possibility of salting-out at low temperatures (Figure 12.6).

Figure 12.6 Simplicity of manufacture and ease of handling are the primary reasons for the popularity of fluid fertilizers produced at rural fertilizer plants such as this one.

12.4.1 Liquid Fertilizers

A liquid fertilizer is a clear solution containing plant nutrients. The principal intermediates used for making liquid NPK fertilizers are urea, UAN (urea-ammonium nitrate) solution, ammonium and orthophosphate or polyphosphate solutions, and finely ground soluble KCl. The liquid fertilizers usually have a brown or green color derived from the wet-process acid used in their preparation.

The most important source of P is ammonium polyphosphate, APP. Ammonia and low-cost, wet-process superphosphoric acid are neutralized in a pipe reactor. The heat produced drives off most of the water, and the high temperature results in formation of polyphosphates in the APP melt. The APP melt is processed into 10-34-0 or 11-37-0 solutions in which 50 to 75% of the P is polyphosphate. The polyphosphate has higher P content and sequesters (holds in solution) both impurities of the acid and any micronutrients that are added. It is a simple mixing process to make liquid NPK fertilizer with APP solution by adding some UAN solution to increase the N content and by adding finely ground KCl to supply K.

The major limitation of liquid fertilizers is their generally lower analysis compared to dry fertilizers. For NP fertilizers, higher concentrations of N and P are produced using UAN solution than using urea. For NPK liquid fertilizers, higher grades are produced using urea rather than UAN solution. Urea produces higher grades because NH_4NO_3 reacts with KCl in solution to produce NH_4Cl and KNO_3. Potassium nitrate has low solubility at low temperature, and crystals form and settle, or the fertilizer salts out. This is more of a problem in the northern states than in the southern. Most of the popular liquid fertilizers have salting-out temperatures of 0°C or lower.

12.4.2 Suspension Fertilizers

The generally low analysis of liquid fertilizers is overcome by the manufacture of suspensions. Suspension fertilizers contain solids held in suspension. The suspended particles may be water-soluble in a saturated solution, insoluble, or both. Higher analysis, lower production cost, and resistance to salting-out are advantages of suspensions compared to liquids. Suspension fertilizers are particularly desirable in regions where high rates of fertilizer K are common. A disadvantage is the need for specialized applicators; most suspension fertilizers must be applied by custom applicators.

Manufacture of suspensions is highly varied, but the process has many parallels to liquid manufacture. Similar materials are used but they may be lower in quality and cost less. The most common suspension agent is attapulgite clay. The usual suspension contains 1 to 2% clay, which forms a gel on vigorous agitation. Particles that settle during storage are easily resuspended when agitated. Suspensions may be stored for weeks or months, but if they are stored too long, large crystals that cannot flow properly through the applicators will grow.

12.5 ADDITION AND INCORPORATION OF MICRONUTRIENTS

Recommended amounts of micronutrients are usually less than 10 kg/ha (about 9 pounds/acre), which makes it difficult to apply them separately and uniformly in the field. For this reason, micronutrients are usually applied in mixed fertilizers rather than separately. Compatibility and uniform distribution are the major objectives of the successful addition of micronutrients to mixed fertilizers.

12.5.1 Addition to Granular Fertilizers

The addition of micronutrients during the manufacture of a granular NPK fertilizer or as one of the intermediates during bulk-blending provides great flexibility, and the cost is low. The micronutrient granule size must match that of the other materials in bulk blends to prevent segregation during mixing, transport, and application. The relative number of micronutrient granules compared to the total number of granules in the fertilizer is very small, which means that the micronutrient granules are widely spaced after application. For example, the use of 8-mesh granules at a rate of 1 kg/ha would provide about one granule per 1,000 cm^2 of the field surface. Some granules would be too far to help some plants and perhaps, depending on material, release too much of the micronutrient to other plants. Even so, bulk-blending of micronutrient intermediates with mixed fertilizers is popular.

Adding micronutrients as a spray or powder during manufacture prior to ammoniation and granulation gives them presence in every granule. The problems of segregation and uniformity of distribution are solved, but the micronutrient intermediate may react unfavorably with other intermediates. Extra precautions are needed to maintain good physical properties of the finished product and keep the micronutrient available. When a chelated micronutrient, ZnEDTA for example, is mixed with superphosphate before ammoniation, acid decomposition of the chelate molecule results in decreased availability of the Zn. When the process is changed by adding the ZnEDTA with the ammoniating solution, the Zn becomes more available.

The major disadvantage of incorporating micronutrients during manufacture is that there is little flexibility, and it is costly to store many grades with a variety of micronutrient contents. If there is a regional deficiency of a nutrient, this method may be the most economical.

Micronutrients are coated to granular products to achieve uniform distribution of the micronutrient throughout the fertilizer. A finely ground micronutrient less than 100-mesh in size is dry-mixed with fertilizer granules, then sprayed with a liquid binder and given additional mixing. Mechanical adhesion or the formation of reaction products on the surface of the fertilizer granules makes the micronutrient coating adhere to the granules. Binders include water, oils, waxes, ammonium polyphosphate, and UAN solutions. Water may be used if it presents no caking hazard. Oil should not be added to NH_4NO_3 because of explosive hazard, and not over 1% oil should be used because the oil may seep through fertilizer bags.

Incorporated and coated micronutrients appear to be similar in agronomic effectiveness. The coatings do not affect the availability of the micronutrents, and provide more flexibility compared to incorporation during manufacture. A disadvantage of coating, compared to incorporation, is higher cost. Micronutrients are not widely coated in the U.S.

12.5.2   Addition to Fluid Fertilizers

The addition of micronutrients to fluid fertilizers has become very popular because the micronutrients can be added just before the fertilizer is applied in the field. The fact that these fertilizers are used immediately allows for great flexibility in manufacture; a wide variety of fertilizers with minimal storage requirements can be produced. The addition of micronutrients to liquid fertilizers, however, is greatly limited by the solubility of the intermediates. Only sodium borate and sodium molybdate have sufficient solubility to be effective when the P source is orthophosphate. Metallic micronutrient sources are more soluble in polyphosphate than orthophosphate in clear fertilizer solutions. Even so, the solubility of salts of Cu, Fe, Mn, and Zn are generally too low to correct severe deficiencies at usual fertilizer application rates. Sequestering agents that react with the micronutrients have been used to prevent their precipitation by phosphate. Ammonium polyphosphate effectively sequesters all of the metallic micronutrients, except Mn. It is easy to custom-mix liquid fertilizers containing a variety of kinds and amounts of micronutrients. A disadvantage of some is that rates of fertilizer application may not allow sufficient amounts of the micronutrient to be applied to correct certain deficiencies.

Micronutrient intermediates are easily added to suspension fertilizers with minimal reaction with other components or other undesirable results. Sulfates of Cu, Fe, Mn, and S in fritted form have been used. Suspensions should be applied soon after preparation to avoid settling difficulties and crystal growth.

12.6 SALT INDEX

Fertilizers are composed for the most part of soluble salts that increase the ionic concentration of the soil solution. This higher ion concentration increases the osmotic pressure of the soil solution and at the same time decreases the water potential. Decreases in water potential are associated with decreases in the rate at which roots and seeds absorb water. Placement of fertilizer close to seeds and roots may reduce germination and growth (Figure 12.7).

The salt index is a measure of the extent to which various fertilizers or intermediates increase the osmotic pressure of the soil solution. Briefly, the salt index is determined by adding fertilizer to soil and incubating for five days. Then, the osmotic pressure of the displaced soil solution is determined. The salt index is a relative value, compared to $NaNO_3$ at 100, and the values of the most widely used materials are given in Table 12.3.

Figure 12.7 Fertilizer placed with or close to the wheat seeds on the left, at a rate comparable to 200 pounds per acre (224 kilograms per hectare), caused a delay in germination.

Table 12.3 Salt Indexes of Fertilizer Materials

Material and grade	Salt index	Index per unit*
Ammonium nitrate, 34-0-0	102	2.90
Ammonium sulfate, 21-0-0	69	3.25
Anhydrous, 82-0-0	47	0.57
Sodium nitrate, 16-0-0	100	6.06
Urea, 45-0-0	73	1.56
Monoammonium phosphate, 11-55-0	27	0.41
Diammonium phosphate, 18-46-0	29	0.45
Superphosphate, 0-20-0	8	0.39
Superphosphate, 0-45-0	10	0.22
Potassium chloride, 0-0-60	116	1.94
Potassium nitrate, 14-0-47	74	1.22
Potassium sulfate, 0-0-54	46	0.85

* Per 20 pounds or 9 kg of $N + P_2O_5 + K_2O$.

Source: Adapted from Rader, White, and Whittaker, 1943.

Nitrogen and K salts have much higher salt indexes than P salts. The cultivar, placement of the fertilizer, time of application, soil properties, and the water content in particular will have some bearing on what effect a fertilizer with a given salt index has. The salt index has its greatest value in selecting fertilizers that will be placed with or near seeds.

12.7 FERTILIZER AND PESTICIDE COMBINATIONS

The application of pesticides with fertilizers has obvious economic advantages. Fewer trips reduce traffic and may reduce soil compaction. The major considerations of combining pesticides with fertilizers are:

1. Appropriateness of timing and placement of the combination.
2. Chemical and physical compatibility.
3. Whether the interaction of pesticide and fertilizer alters their effectiveness.
4. Metal corrosion and deterioration of hoses and other equipment.

12.7.1 Herbicide and Fertilizer Combination

Since the early 1970s, the popularity of herbicide and fertilizer combinations has fostered the *weed and feed* concept. One of the most important considerations is choosing the time that the herbicide-fertilizer combination should be applied for the greatest effectiveness of each. For many crops it is customary to split the N application, which often means that the time of application for both N and herbicide is at preplanting, planting, or post-emergence. The most popular combinations are N solutions, UAN, and herbicides such as fluometuron for cotton and atrazine or simazine for corn. Nitrogen solution plus 2,4-D is used to control weeds in small grain, sorghum, and pastures. The N readily moves into the soil with water, and the surface application of the herbicide is effective. The larger quantity of fluid may result in more uniform application and render the herbicide more effective. An application of 140 to 190 L/ha of UAN used as a carrier results in a N application of 50 to 68 kg/ha (45-61 pounds/acre).

Post-emergence application causes leaf burn unless drop nozzles are used to direct the solution below the leaves. Burn injury increases with rate of application and advanced stage of plant growth. Growers may express concern about leaf burn. If the leaf burn occurs early in the season, yield may not decrease significantly. It has been reported that as much as 134 kg/ha (120 pounds/acre) of N has been applied using UAN on four-leaf corn without reducing corn yields. Application after the six-leaf stage, however, should be regarded as a salvage operation.

Problems of incorporation of herbicides with fluid fertilizers include precipitation of salts, formation of thixotrophic gels, and increase in salting-out temperature of liquids. The herbicide should be mixed just before application to minimize problems. It should be continuously agitated. A simple field test can be made to obtain some visible evidence of compatibility:

1. Place a liter or pint of the fluid fertilizer in a jar.
2. Add an amount of the herbicide to equal the proportion of the formulation.
3. Shake the jar and immediately observe separation or gelling. Repeat observation in 30 minutes.

If an emulsifier is to be used, the test can be used to observe the effectiveness of adding the emulsifier.

Some changes are not visible and may only appear when the effectiveness of the herbicide is measured. Many herbicides are complex organic molecules that have many reaction sites for adsorption and reaction with fertilizer salts. Herbicidal action may be increased or decreased. Attapulgite clay in suspension adsorbs paraquat and renders it ineffective.

Herbicides are also incorporated into dry fertilizers. Dinitroanilines and carbamates are sprayed on or impregnated into the fertilizer just before application. These combinations are spread on the soil and incorporated by subsequent tillage. Adding dry granular herbicides to granular fertilizers creates some of the same segregation and reaction problems encountered in adding granular micronutrients to granular fertilizers. In addition, the effectiveness of the herbicide is likely to be reduced in prolonged storage of granular combinations.

12.7.2 Insecticide and Fertilizer Combination

The combination of fluid fertilizers with herbicides has been much more successful than their combination with insecticides. Generally, insecticides are applied to the foliage and fertilizers are applied to the soil. One application, although not very popular, is the use of a liquid starter fertilizer-insecticide combination to control corn rootworm. Some problems encountered include compatibility of materials, phytotoxicity, and insecticide performance.

REFERENCES

Barber, S. A., R. D. Munson and W. B. Dancy. 1985. Production, Marketing and Use of Potassium Fertilizers. *Fertilizer Technology and Use.* Soil Sci. Soc. Am., Madison, WI.

Engelstad, O.P., Ed. 1985. *Fertilizer Technology and Use.* 3rd ed. Soil Sci. Soc. Am., Madison, WI.

Follett, R. H., L. S. Murphy and R. L. Donahue. 1981. *Fertilizers and Soil Amendments.* Prentice Hall, Englewood Cliffs, NJ.

Hoffmeister, G. 1979. *Physical Properties of Fertilizers and Methods for Measuring Them.* Bull. Y-147. TVA, Muscle Shoals, AL.

International Fertilizer Development Center. 1979. *Fertilizer Manual.* TVA, Muscle Shoals, AL.

Jackson, J. E. and G. W. Burton. 1958. An Evaluation of Granite Meal as a Source of Potassium for Coastal Bermudagrass. *Agronomy J.* 50:307-309.

Jones, U. S. 1982. *Soil Fertility and Fertilizers.* Prentice-Hall, Reston, VA.

National Fertilizer Development Center. 1983. *Fertilizer Trends 1982*. TVA, Muscle
    Shoals, AL.

Pierre, W. H. 1933. Determination of Equivalent Acidity and Basicity of Fertilizers.
    Industrial and Eng. Chem. 5:229-234.

Rader, L. F., L. M. White and C. W. Whittaker. 1943. The Salt Index: A Measure of the
    Effect of Fertilizers on the Concentration of the Soil Solution. *Soil Sci.* 55:201-208.

Tisdale, S. L., W. L. Nelson and J. D. Beaton. 1985. *Soil Fertility and Fertilizers*. 4th ed.
    Macmillan, New York.

Whittaker, C. W. 1948. *Ammonium Nitrate for Crop Production*. Cir. 171, USDA, Wash-
    ington, D.C.

Young. R. D. 1983. *Mixed Fertilizers*. Mimeo. TVA, Muscle Shoals, AL.

CHAPTER **13**

Soil Fertility Evaluation

An evaluation of soil fertility is based on observations and tests of both plants and soils. Two primary purposes of the tests are to make fertilizer recommendations and to measure the effectiveness of fertilizer practices. Soil tests can be carried out before crops are planted and are the basis of fertilizer recommendations for most annual field and vegetable crops. Plant analysis is a popular way to make recommendations for perennial crops, because samples of tissue are collected after crops are established and there is a long time after testing for nutrient accumulation and growth. In addition, the roots of perennials, tree crops for example, are permanent residents throughout a very large volume of soil, which makes soil sampling difficult. Both kinds of tests and plant-deficiency symptoms are used for diagnostic purposes.

13.1 PLANT-DEFICIENCY SYMPTOMS

Plants, like animals, exhibit unique symptoms to various nutrient deficiencies. Light green- or yellow-colored leaves and slow growth are common nitrogen-deficiency symptoms of lawns. Nitrogen, P, K, and S are mobile in plants and are readily moved from the older and lower leaves to the upper or newly formed parts of plants when a deficiency occurs (Figure 13.1). In fact, it is normal for mobile nutrients to accumulate in the vegetative parts and be moved to the fruit before harvest time. As N is removed from the leaves of corn for grain development late in the season, yellowing and death of the lower leaves is normal. Deficiency symptoms of mobile nutrients appear first on the lower leaves or oldest tissues. By contrast B, Ca, Cu, Fe, and to a lesser degree, Mn, are quite immobile in plants and deficiency symptoms appear first on the most recently formed leaves or tissues.

Deficiency symptoms are useful to diagnose plant growth problems and in selecting sites for soil fertility experiments. If detected early enough in the season, deficiency symptoms can be considered when deciding whether an additional amount of fertilizer should be applied. Sometimes deficiency symptoms appear

Figure 13.1 Nitrogen deficiency symptom of corn (maize) is a yellow mid-rib of the lower leaves. Eventually, the entire leaf may turn yellow and die.

early in the season and disappear, with no apparent reduction in yield. On the other hand, certain deficiency symptoms may appear in mid-season or later and may be associated with yield reductions as large as 50%.

13.2 SOIL TESTS

The tests for soil pH and lime requirement and the recommendations for liming are considered in Chapters 4 and 5. Here, the emphasis is on the use of tests and procedures designed to make fertilizer recommendations.

The primary objective of the use of fertilizer is to make a profit. The economical use of fertilizer depends on knowledge that can relate the use of fertilizer to a predicted increase in yield. Soils can be sampled and soil tested, and fertilizer recommendations developed before the growing season begins. This is of particular importance to crops which have only about 100 days to grow and accumulate nutrients. Since yields are also a function of climate and management, the effects of climate and management must also be integrated into fertilizer recommendations. A soil-testing program designed to make fertilizer recommendations typically has four parts: (1) sample collection, (2) chemical analysis, (3) interpretation of the chemical tests, and (4) recommendations based on the tests for the crops to be grown.

13.2.1 Soil Sampling

Most soil tests use only 0.1 to 5 grams of soil, yet the volume of soil represented will likely exceed 10,000 tons. The ability to obtain a truly representative sample still remains one of the weakest links in evaluation of soil fertility. Instructions for sampling should be obtained from the soil-testing laboratory. The laboratory will supply the farmer with field information sheets and soil sample containers as well as the proper instructions. Four things are needed for sampling: a sampling tool (soil sampling tube, auger, or shovel); a clean pail (plastic will reduce the likelihood of micronutrient contamination); a field information sheet; and soil sample containers.

Before obtaining a sample, evaluate the field for uniformity of soil characteristics and past management practices. A sampling unit or area should be quite uniform in those characteristics that affect yields. The size of the sampling unit may vary but generally should be field size or sufficiently large so that it can be fertilized differently from other areas. Avoid sampling small areas of unusual drainage, areas that have undergone unusual management in the past, or sites where manure or lime piles had been located. The sampling area should be sketched on the field information sheet and a label put on the containers that will identify the samples and associate them with the field. Unusual areas may be sampled separately if the farmer so desires. The sample should represent the plow layer where conventional tillage is used. From the sampling area, obtain 20 to 30 cores that are uniform in size and represent the same soil depth. Obtain the cores at random as shown in Figure 13.2.

In conventional tillage systems, the sampling depth should be the plow depth. Fertilizer carry-over and build-up of fertility means that samples containing soil from below the plow layer will tend to be biased to lower tests for P and K. As shown in Chapter 7, the P content of the plow layers of many fields is now so high that P fertilizers are not recommended. No-till systems have created P soil test gradients in the upper 15 cm of soils, as compared to conventional tillage. Higher P concentrations develop at the soil surface because of limited mobility of surface applied P and the deposition of P in the organic residues returned to the surface of the soil. Acidity produced by N fertilizer use has likewise created pH gradients in the upper few cm of soil. As a consequence, some laboratories recommend a sampling depth of 0 to 7.5 cm for no-till and disk-tilled fields. In Georgia, a 10-cm sampling depth is recommended for no-till fields, pastures, lawns, and turf. It may be advisable to sample two depths, 0 to 5 cm and 5 to 20 cm, for no-till fields. Additional research is needed to develop better sampling techniques and fertilizer and lime recommendations for no-till fields.

Nitrate and sulfate sulfur tend to be quite mobile in soils, and deeper sampling is required in order to test them. Several Great Plains states recommend sampling at 60 to 120 cm in testing for the available N (nitrate). Sampling to 60 cm is

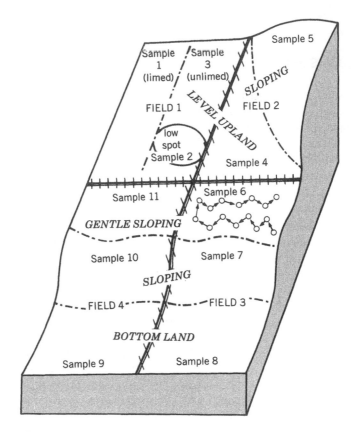

Figure 13.2 Example of how to lay out a field for soil sampling.

encouraged in the Canadian prairie provinces for testing of nitrate, sulfate, and soil salinity.

Place the cores in a clean pail and thoroughly mix. Fill the soil sample container and send sample and field information sheets to the soil-testing laboratory. If a nitrate test is requested, moist soil samples should be quickly dried without the use of artificial heat before samples are shipped. This will minimize any change in nitrates through further microbiological activity. Allowing samples to air-dry slowly is generally acceptable because most laboratories dry and pulverize the samples before testing. In any case, it is important to follow the instructions of the testing laboratory because their interpretation of soil test results is based on prescribed handling of samples.

Seasonal trends in soil test results have been reported. The seasonal trend has been the most consistent for K. Soil samples taken during the growing season will often test lower than samples taken in the winter or early spring. Illinois adjusts K soil-test results on samples taken between September 30 and May 1. In most cases, the seasonal trends are small, and for routine soil tests, sampling is encouraged whenever it can be done. Most soil-test laboratories recommend sampling and testing be done every two to four years.

13.2.2 Types of Tests

The kinds of tests available depend on local soil and crop conditions. A standard test usually includes available P, exchangeable K, Ca, and Mg. It also includes their saturation percentages, the cation exchange capacity (CEC), pH and lime requirement. Greenhouse tests usually include, in addition, salinity and nitrate nitrogen. Some laboratories test for organic matter content, salinity, sulfate, and certain micronutrients and heavy metals. Some laboratories also offer tests for water quality.

13.2.3 Soil Test Correlation

A good soil test for a nutrient will effectively measure a quantity that will closely correlate with the amount of the nutrient that plants will take up, thus it will also correlate well with yields. This soil fraction of the nutrient is typically referred to as the available fraction. Greenhouse experiments can be used to do much of the preliminary work in selecting the most efficient soil test method. The development of a soil test correlation program requires field experiments that accomplish four things: (1) experimental treatments must be replicated; (2) treatments must include rates of application for a given nutrient element to determine the quantity of fertilizer required for maximum yield at a given soil test level; (3) other growth factors should be managed to obtain high yields, including an adequate supply of other nutrients; (4) the trials must be repeated at sufficient locations and number of years to estimate variability caused by climate and location (soil variability). The data from the experiments are used to construct figures, such as Figure 13.3, which relate soil tests to yields. The use of relative yields rather than actual yields enables all trials to be shown as one figure.

The data in Figure 13.3 show that cotton requires a higher K soil test for a given relative yield than does corn, whereas soybeans tend to require an intermediate amount. Similar statements are commonly made for P. For instance, small grains, alfalfa, and clover require higher soil tests than soybeans or corn for the same relative yield. Moreover, because of soil differences, including mineralogy, a K test of 20 for soils in group 1 is as effective as a test of 40 for soils in group 3. The soil groupings in this example are based on CEC, groups 1, 2, and 3 having CEC of 10, 20, and 30 meq/100 g, respectively. These results from Alabama parallel findings in other states, namely, that finer-textured soils must have a higher K test to produce the same crop yield as compared to sandy soils. In Ohio, the desired or optimum K test in pounds per acre is 220 plus 5 times the CEC. This formula works out to 270, 320, and 370 for soils with CEC of 10, 20, and 30 meq/100 g. In Georgia, soils are grouped into three categories for making K recommendations, including (1) coastal plain soils; (2) piedmont, mountain, and limestone valley soils; and (3) soils from landscapes, golf greens, greenhouses, and flower beds. Similar studies are conducted to correlate soil P tests with yields.

Soil test correlation data are converted into descriptive terms and into useful interpretations as shown in Table 13.1. Many different soil test levels are differ-

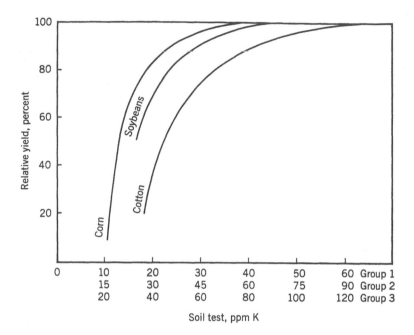

Figure 13.3 Correlation between potassium soil test and the relative yields of crops grown on three soil groups in Alabama. (Adapted from Cope and Rouse, 1973.)

Table 13.1 Soil Test Rating, Relative Yield, and Recommendations

Rating	Relative yield, percent	Recommendations
Very low	<50	Large applications for soil building.
Low	50–75	Annual applications to produce maximum yield and increase soil fertility.
Medium	75–100	Normal applications to produce maximum yields.
High	100	Small applications to maintain soil fertility.
Very high	100	None until soil tests drops to high range.
Extremely high	100	P may be excessive, contaminating the environment, and cause Fe or Zn deficiency.

Source: Adapted from Cope and Rouse, 1973.

entiated, depending on need and situation. In Table 13.1, a soil test value associated with a relative yield of less than 50% of the maximum is considered very low. Crops grown on soils with a very low soil test are predicted to yield in the range of 0 to 50% of the maximum yield for the soil and crop conditions. Such soils may be treated with large amounts of fertilizer to build up soil fertility. Soil tests of low and medium rating represent soils that are predicted to yield 50 to 75% and 75 to 100% of maximum yield. Soils that test high are expected to produce the maximum yield without fertilizer. Very high and extremely high levels indicate that soils contain more of the nutrient in available form than is needed to produce the maximum yield. Such levels are not uncommon results for garden samples. Extremely high soil tests may be associated with nutrient toxicities, nutrient imbalances, or environmental contamination. Thus, soil test correlation experiments provide the information to correlate soil tests of various soils with the relative yield expected for various crops. The next step in making a specific fertilizer recommendation is to calibrate soil tests with yields produced by varying amounts of fertilizer.

13.2.4 Calibration of Soil Tests with Yield Responses

Soil test calibration experiments measure crop response to varying rates of a fertilizer nutrient applied to plots with various soil test levels. These experiments provide the data needed to construct graphs of the nature of Figure 13.4. The lower the soil test, the greater is the amount of fertilizer needed to produce a particular yield or the maximum yield. Knowing the cost of fertilizer and the price of the crop permits a calculation and prediction of the amount of fertilizer that will produce the yield that will in turn earn the maximum profit.

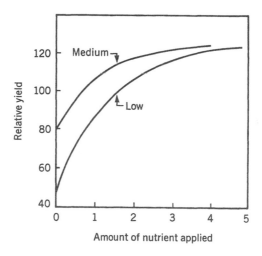

Figure 13.4 Typical yield responses from nutrient additions for medium and low soil test levels.

In practice, farmers are asked to record their yield goal on the field information sheet. This value is based in part on weather expectations, previous and current management practices, and soil conditions. Then soil calibration data are used to recommend the amount of fertilizer that is predicted to produce the yield goal. Theoretically, maximum profit is earned with yields that are less than the maximum, usually in the 90 to 95% of the maximum range.

The curves in Figure 13.5 represent four different fertilizer situations. Curve A is an adverse situation where yields without fertilizer are low, and little potential exists for increasing the yield with fertilizer. For example, this may be a very droughty sand soil. The fertilizer optimum rate is less than two units. Curves B, C, and D represent situations with increasing yield potential; differences in yield

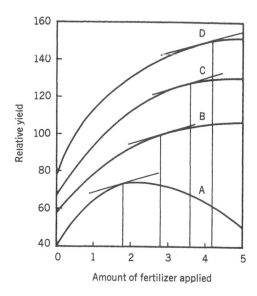

Figure 13.5 Effect of potential yield of four different soils on yield response to fertilizer.

potentials could reflect differences in soil characteristics, management, or weather. For curve C, the maximum profit or optimum yield is obtained with the use of 3.6 units of fertilizer. There is a tendency to recommend a fertilizer rate that will produce a yield near the maximum, because with underfertilization, loss of profit is greater than with overfertilization as shown in Table 13.2.

Table 13.2 Effect of Underfertilizing versus Overfertilizing on Net Return

Fertilizer	Rate	Yield value	Cost of nutrients	Net return	Difference from optimum
One-fourth less	2.7	120.5	8.1	46.4	−4.3
Optimum	3.6	127.5	10.8	50.7	0
One-fourth more	4.5	129.0	13.5	49.5	−1.2

Source: Adapted from Barber, 1973.

13.2.5 Subsoil Testing

It is generally accepted that subsoil water and nutrients are important for plant growth. Even so, very few quantitative studies of the importance of subsoil fertility have been made. These studies suggest that from 0 to 80% of the total nutrient uptake occurred in the subsoil. In the U.S., it is routine in areas with ustic soil moisture regime to test for subsoil nitrate and use the test results for making N fertilizer recommendations.

The importance of subsoil fertility is related to rooting habit of plants, the level of subsoil fertility, and the moisture conditions of the subsoil. If the entire soil profile is kept moist and at a uniform level of fertility, root growth and nutrient uptake will be much greater in the upper soil layers than the deeper soil layers. When the surface soil dries, nutrient uptake tends to shift to lower soil layers. It has been reported that in Nebraska the first cutting of alfalfa absorbed only 5% of the P from deep in the subsoil. The third cutting, produced after the surface soil had become droughty, absorbed 62% of the P from deep subsoil layers. A recent experiment in Kansas was designed to measure P uptake by sorghum grown in plastic cylinders. As the level of available P in the top soil decreased, more P was taken up from the subsoil. In fact, when the level P in the surface soil was low and that of the subsoil high, most of the P was absorbed from the subsoil. The experimenters found a close relationship between P uptake and the available P level and root density. It has been observed that the P fertilizer needs of alfalfa decrease when roots become established in soils with high levels of P in the subsoil.

At present, little subsoil testing is done, although it can be an important diagnostic tool. However, routine subsoil testing of some soils is recommended for making K recommendations in Delaware. The soils are Ultisols with very siliceous and sandy plow layers that are underlain with clayey argillic (or kandic) horizons with considerable K. Perhaps, in the future, more effort will be directed toward greater utilization of subsoil fertility, by altering plants genetically so that they tap the subsoil. Then subsoil testing will become more important.

13.3 PLANT ANALYSIS

The nutrient concentration of a plant is an integrated value that reflects all the factors that have influenced nutrient concentration up to the time of sampling. The nutrient concentration is related to growth and changes over time, as shown in Figure 13.6. When a nutrient is deficient for maximum growth under most conditions encountered in the field, an increase in the supply increases both growth or yield and the nutrient's percentage of composition in the plant. Additional uptake of a nutrient beyond the point of maximum yield continues to increase the nutrient concentration, into the zone of luxury consumption, with no yield increase. Eventually, the concentration in the plant may become toxic and cause yields to decline.

13.3.1 Plant Composition and Yield

The relation between plant composition and yield or growth is the basis for information that is gained through plant analysis and used to make fertilizer recommendations, as shown in Figure 13.7. The nutrient concentration at the maximum yield is the critical nutrient concentration (CNC); there is a transition zone in which the nutrient concentration changes rather abruptly from deficient to adequate or sufficient. Much research has been done to establish the CNC and transition zones for many plants, and plant analysis is commonly used to predict the potential for applying fertilizer to increase yield relative to the maximum

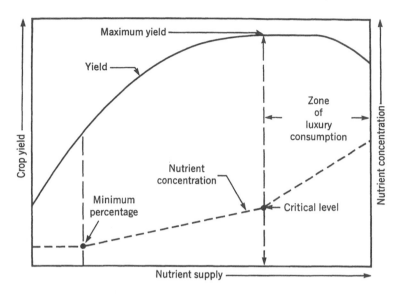

Figure 13.6 Relation between nutrient supply, crop yield, and nutrient concentration in plants. (After Brown, 1970.)

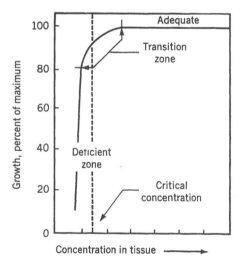

Figure 13.7 Relation between the concentration of a nutrient in the plant, and its deficiency and adequacy for maximum growth. (Adapted from Ulrich and Hills, 1967.)

yield. Some values representative of deficiency and sufficiency for a few crops commonly grown in the western part of U.S. are given in Table 13.3.

Fruit and forest trees grow over a long time period and CNC values can be of great use in diagnosing problems and formulating management practices. In Michigan, plant analysis was used to document the fact that there were virtually no micronutrient shortages, that in fruit orchards there had been widespread use of complete fertilizers, which were 90% unnecessary. Measuring changes in plant

Table 13.3 Plant Tissue Analysis Guide for Some Western Crops

| Crop | Sampling date | Nutrient* | Nutrient level | |
			Deficient	Sufficient
Alfalfa	One-tenth bloom, middle	N	—	—
	half of stem	P	500	800
		K	0.7	0.8
Cantaloupe	Early fruit, petiole of	N	5,000	9,000
	sixth leaf from vine tip	P	1,500	2,500
		K	3.0	5.0
Lettuce	Heading, mid-rib of	N	4,000	8,000
	wrapper leaf	P	2,000	4,000
		K	2.0	4.0
Potatoes	Mid-season, petiole of	N	6,000	9,000
	fourth leaf from growing	P	8,000	16,000
	tip	K	7.0	9.0
Tomato (canning)	Early bloom, petiole	N	8,000	12,000
	of fourth leaf from	P	2,000	8,000
	growing tip	K	3.0	9.0
Watermelon	Early fruit, petiole of	N	5,000	9,000
	sixth leaf from	P	1,500	2,500
	growing tip	K	3.0	5.0

* Nitrogen as nitrate N; P as phosphates soluble in acetic acid, in parts per million; and K as percentage of total K.

Source: Data from Western Fertilizer Handbook, 1985, Soil Improvement Comm. Cal. Fert. Assoc.

composition is a standard method to evaluate the effectiveness of fertilization practices. When tissue analysis is used to measure a forest's response to fertilization, it may take several years before measurable differences appear in foliar dimensions, tree height and diameter, and wood volume.

In general, a nutrient's percentage of composition in a plant decreases as plants age and there is great variation between different parts of the plant. Research data for most economic crops has been summarized and placed in tables, showing when and which part of the plant to sample. For example, the CNC values for corn have been given as 3% for N, 0.03% for P, and 2% for K in the leaf opposite and below the uppermost ear (ear leaf) at silking time (see Table 13.3).

13.3.2 Crop Logging

Crop logging, which consist of periodically measuring the plant composition, has been widely used for high-value crops, including sugar cane, pineapples, and sugar beets. For crops that grow over a long period of time, unusual heavy rainfall or reducing conditions may greatly alter the available supply of some nutrients. Maintaining an adequate supply of N for plants over a long growing season without potential contamination of water supplies is a frequent problem. Periodic plant analysis or crop logging was used in California to determine N fertilization practices that would maximize sugar production and minimize water pollution.

For maximum sugar production by sugar beets, it is desirable to have a limited or small amount of N available early in the season to encourage high sugar content of roots. Later in the season it is desirable to have abundant N to increase the quantity of roots. In an experiment, 90 kg N/ha were applied at thinning time, and petiole N remained above the sufficient level until the latter part of July, when an additional application was made (Figure 13.8). This method produced more sugar than if all of the N had been applied at planting time and minimized the potential for water pollution.

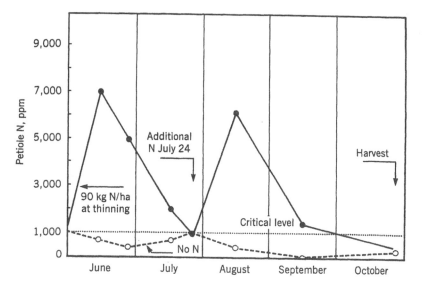

Figure 13.8 Nitrate content in petioles of sugar beets during the growing season in response to two applications of nitrogen fertilizer and no nitrogen application. (Modified from Ulrich and Hills, 1967.)

13.3.3 Rapid Plant Tissue Tests

Analytical plant analysis is an indispensable research tool, one which is useful in soil management, but the tests are expensive and time consuming. Inexpensive test kits for N, P, and K are available and have been widely used to diagnose plant growth problems in the field. They are particularly useful to verify a particular deficiency symptom. They are of two types: selected tissue is cut up into small pieces and placed in a vial with distilled water, or plant sap is squeezed onto a prepared paper. Chemicals are used to assay the nutrient content, which, in essence, is a measure of nutrients in plant sap. Particular plant parts are sampled at various stages of growth, the same as in analytical tests.

Since N deficiency is associated with the loss of chlorophyll, a "chlorophyll" meter has been developed and the readings from intact leaves by the meter found to correlate with yield response to N. This offers a rapid, in-field method of determining if N fertilization is needed while the crop is growing.

13.3.4 Diagnosis and Recommendation Integrated System

An alternate method of evaluating crop nutrient status through the use of tissue analysis was developed by Beaufils (1973), called the diagnosis and recommendation integrated system (DRIS). This system incorporates tissue analysis with other yield parameters in making recommendations.

Beaufils proposed that the DRIS methology should follow four steps:

1. Define the characters to be improved and all factors that are suspected to affect those characters.
2. Gather all reliable data available from field operations and experiments.
3. Relate yield and all external parameters (i.e., light, rainfall, etc.)
4. Relate yield and all internal parameters (i.e., nutrient content).

The DRIS system emphasizes the importance of nutrient balance within plant tissue and states that maximum yield may never be obtained unless the proper balance or ratio of nutrients is maintained in plant tissue. Having this proper ratio of all plant nutrients does not in fact guarantee a maximum yield, because the crop could be destroyed by hail, flooding, or other causes. If the plant nutrient ratio is incorrect, the yields will be lower.

The DRIS system has an advantage over trying to determine critical values in that nutrient ratios in plant tissue tend to be rather constant throughout the growing season. Thus, by applying this system, the farmer avoids having to sample the crop at a precise growth stage and estimate the change that will occur during the rest of the period.

For successful use, the DRIS system must have known *norms* or indices that are associated with the maximum yields for each crop. These have been developed by utilizing a large data base collected from the literature from each crop. But it has been shown that a large data base is not essential to establish DRIS norms if a small data base can be collected for very high yielding populations.

The method of calculation of the norms must be readily obtainable. Norms have been developed for many of the agronomic crops, including corn, soybeans, wheat, sugar cane, potatoes, and sorghum. Norms have also been developed for fruit trees, forage crops, grasses, and forest species. Letzsch and Sumner (1983) have developed a computer program for calculation of the norms and offer copies of it for a nominal cost.

REFERENCES

Barber, S. A. 1973. The Changing Philosophy of Soil Test Interpretations. In *Soil Testing and Plant Analysis*. Soil Sci. Soc. Am., Madison, WI.

Beaufils, E. R. 1973. *Diagnosis and Recommendation Integrated System (DRIS)*. Dept. Soil Sci. and Agronomy. Univ. Natal, Pietermaritzburg, South Africa.

Brown, J. R. 1970. *Plant Analysis*. Bull. SB881, Missouri Ag. Exp. Sta., Columbia, MO.

Comerford, N. B., G. Kidder and A. V. Mollitor. 1984. Importance of Subsoil Fertility to Forest and Nonforest Plant Nutrition. In *Forest Soils and Treatment Impacts*. Dept. Forestry, Wildlife and Fisheries, University Tennessee, Knoxville, TN.

Cope, J. T. and R. D. Rouse. 1973. Interpretation of Soil Test Results. In *Soil Testing and Plant Analysis*. Soil Sci. Soc. Am., Madison, WI.

Ellis, B. G. and R. A. Olson. 1986. *Economic, Agronomic and Environmental Implications of Fertilizer Recommendations*. North Central Regional Research Pub. 310, Mich. Sta. Agr. Exp. Sta., East Lansing, MI.

Fox, R. L. and R. C. Lipps. 1964. A Comparison of Stable Strontium and ^{32}P as Tracers for Estimating Alfalfa Root Activity. *Plant and Soil* 20:337–350.

Gilliam, J. W., T. J. Logan and F. E. Broadbent. 1985. Fertilizer Use in Relation to Environment. In *Fertilizer Technology and Use*. 3rd ed. O. P. Engelstad, Ed. Soil Sci. Soc. Am., Madison, WI.

Goni, C. 1994. Preliminary DRIS Norms in Highbush Blueberry (*Vaccinum corymbosum L.*) M.S. thesis, Michigan State University.

Kenworthy, A. L. 1973. Leaf Analysis as an Aid in Fertilizing Orchards. *Soil Testing and Plant Analysis*. Soil Sci. Soc. Am., Madison, WI.

Leaf, A. L. 1973. Plant Analysis as an Aid in Fertilizing Forests. *Soil Testing and Plant Analysis*. Soil Sci. Soc. Am., Madison, WI.

Letzsch, W. S. and M. E. Sumner. 1983. Computer Program for Calculating DRIS Indices. *Commun. in Soil Sci. Plant Anal.* 14:811–815.

Melsted, S. W. and T. R. Peck. 1973. The Principles of Soil Testing. *Soil Testing and Plant Analysis*. Soil Sci. Soc. Am., Madison, WI.

Munson, R. D. and W. L. Nelson. 1973. Principles and Practices in Plant Analysis. *Soil Testing and Plant Analysis*. Soil Sci. Soc. Am., Madison, WI.

Peck, T. R. and S. W. Melsted. 1973. Field Sampling for Soil Testing. *Soil Testing and Plant Analysis*. Soil Sci. Soc. Am., Madison, WI.

Pickielek, W. P. and R. H. Fox. 1992. Use of a chlorophyll meter to predict sidedress nitrogen requirements for maize. *Agron. J.* 84:59–65.

Pothuluri, J. V., D. E. Kissel, D. A. Whitney and S. J. Thien. 1986. Phosphorus Uptake from Soil Layers Having Different Soil Test Phosphorus Levels. *Agron. J.* 78:991–994.

Schepers, J. S., D. D. Francis and C. Clausen. 1990. In *Agronomy Abstracts*, p. 280. ASA, Madison, WI.

Soil Improvement Comm. Cal. Fert. Assoc. 1985. *Western Fertilizer Handbook*. 7th ed. Interstate, Danville, IL.

Sumner, M. E., R. B. Reneau, Jr., E. E. Schulte and J. O. Arogun. 1983. Foliar Diagnostic Norms for Sorghum. *Commun. in Soil Sci. Plant Anal.* 14:817–825.

Ulrich, A. and F. J. Hills. 1973. Plant Analysis as an Aid in Fertilizing Sugar Crops. *Soil Testing and Plant Analysis*. Soil Sci. Soc. Am., Madison, WI.

Walworth, J. L., H. J. Woodard and M. E. Sumner. 1988. Generation of corn tissue norms from a small high-yield data base. *Commun. Soil Sci. Plant Anal.* 19:563-577.

Whitney, D. A., J. T. Cope and L. F. Welch. 1985. Prescribing Soil and Crop Nutrient Needs. *Fertilizer Technology and Use*. 3rd ed. Soil Sci. Soc. Am., Madison, WI.

CHAPTER 14

Application and Use of Fertilizers

In general, about 50% of the N, 20% of the P, and 35% of the K, or less, in fertilizers is absorbed by crops during the year of application. Achieving the maximum biological efficiency in using fertilizers depends on time of application and their placement. Factors that influence decisions about when to apply a fertilizer and where to place it include nutrient mobility and fixation in the soil; plant needs and the location of roots; losses by volatilization and leaching; the amount of soil moisture and temperature of soils; and salt effects on germination. In general, the closer the time of application matches plant nutrient demands, decreases the opportunity for fixation and loss, and increases the opportunity for absorption and a yield response to the nutrient applied. Rates of application and use are also affected by economic and environmental considerations and convenience.

14.1 FERTILIZER APPLICATION PRACTICES

Fertilizers are placed either on or in the soil. The major methods of surface placement are broadcast of dry materials, spraying of liquids, and application in irrigation water. Methods of placing fertilizer in the soil includes band placement, plow down, injection, and application in irrigation water. Only a very minor amount of fertilizer is applied to leaves as a foliar spray.

Fertilizer placement and time of application are importantly affected by the practices used to establish and manage a crop and the properties of soils and fertilizers. In general, dry and liquid forms are of equal value if similarly placed in the soil and applied at the same period of time. The major factor is amount of nutrient applied. Nutrients from both dry and liquid forms must appear in the soil solution as ions before they can be taken up by plants. As the level of soil fertility increases, placement of fertilizer becomes less important.

The major concern in applying fertilizers is that the nutrients be used efficiently without any detrimental effects on plants or the environment. The major consideration in the placement and time of applying N fertilizers is to prevent the loss of N through volatilization, leaching, and denitrification. Fixation of P, the immobility of P, and the limited mobility of K in soils are the major considerations in the placement and timing of application of P and K fertilizers. In some soils, K fixation becomes an important consideration. There is an interaction between placement and time of application, and it is desirable to discuss them together.

14.1.1   Injection of Anhydrous Ammonia

The method of application of NH_3 is related to its gaseous nature. Anhydrous ammonia, 82-0-0, has a vapor pressure at 0°C of 47 pounds per square inch of gauge pressure, and at 38°C of 197. Because it is a volatile liquid, it must be placed 15 to 25 cm deep in the soil to minimize loss by volatilization. As shown in Figure 11.4, the amount lost is related to soil type, moisture content, and depth of application. For row crops, the farmer usually chooses an applicator knife or injector to apply the fertilizer between alternate pairs of rows, and pesticides may be applied at the same time. For forages and small grains, the injectors need to be about 30 to 40 cm apart to obtain fairly uniform distribution of the N for the roots. Application of part of the N during the growing season ensures more efficient use than if all of the N is applied before or at planting time. Now for row crop production it is popular to combine the application of NH_3 with cultivation or with application of a herbicide.

It has become popular to apply anhydrous ammonia with other tillage operations, especially prior to planting. On the other hand in the Great Plains, where loss of N by leaching is minimal, delivery lines place anhydrous ammonia at the bottom of the tilled layer of undercutting sweeps. This practice after grain harvest effectively controls weeds and leaves crop residues on the surface to reduce soil erosion. In humid regions where nitrate leaching losses are likely, the application should be delayed until soil temperature is below 18°C (50°F) in the fall. The use of nitrification inhibitors to reduce nitrification of fall-applied ammonium-N is unreliable. But some benefit may be gained when a nitrification inhibitor is applied with the injection of NH_3 (and the plow down of other N carriers) in the fall on fine-textured soils. Frequently, the fall is a time of low labor demand. In some situations, soils are dry and firm and the price of the N may be discounted. Fall application of N in the humid region, however, is losing favor because of concern about nitrate pollution of groundwater.

In the U.S., the first injection of anhydrous ammonia into the soil occurred in the Mississippi Delta region in the early 1940s. The advantages of placement of P and K fertilizer within the soil may encourage future farmers to combine the injection of NH_3 with the injection of P and K.

14.1.2   Band Application

During the early days of using fertilizer, agricultural soils were more infertile than today, fertilizers were relatively more expensive, and low rates of application were common. To obtain maximum effectiveness, farmers applied the fertilizers in bands along with the seed. Fertilizers became more concentrated over time, and the salt index and rates of application increased. The salt affected germination, and fertilizer applicators that separated seed and fertilizer were introduced. The standard recommendation for row crops is a band 2 inches to the side and 2 to 6 inches below the seeds to make sure seed and fertilizer are separated enough to avoid salt injury. In this location, the fertilizer is usually in moist soil where early root growth can provide the plant with quick access to the fertilizer. Band placement reduces P fixation and puts a readily available source of nutrients close to roots, allowing the crops to get off to a fast and vigorous start. There is less contact between the P fertilizer and soil when banded as compared to broadcast, creating zones or small areas of much higher P availability. Root growth in a fertilized soil band or zone has commonly been observed to be greater than root growth in the surrounding unfertilized soil (Figure 14.1). This effect is due mainly to the P.

Figure 14.1 Diffusion of nutrients from fertilizer granules into the surrounding soil enriched the soil and caused greater growth of the corn roots.

Band placement of N is not effective. Nitrate is not fixed or adsorbed by the soil and remains very mobile. Surface application is as good as placement in the soil if there is rain or irrigation to move the N into the soil. Ammonium in fertilizer is usually rapidly nitrified to NO_3^- and becomes very mobile. Now fertility levels for P and K are frequently so high that band placement of P or K does not increase yield more than does broadcast application. However, plant analysis may show slightly greater uptake of fertilizer P and K early in the growing season when it is band placed.

As indicated, there is less and less benefit from banding fertilizer as soil fertility levels increase. There are, however, some important situations in which band placement is effective. Yield responses with band application of P on Oxisols with high P-fixing capacity are common. Band application of K is also effective on soils with unusually high K-fixing capacity. Band application helps to increase the uptake of micronutrients from fertilizers that are subject to rapid soil fixation in all soils. Low spring temperatures in the soil appear to be the reason for the response of spring wheat to banded P on the northern Great Plains.

14.1.3 Broadcast Application

Broadcast application is the spreading of the fertilizer uniformly over the entire soil surface. It is the only practical way to fertilize lawns, forage, and pasture fields, and forests on mountain slopes. As already indicated, anhydrous ammonia is not suitable for broadcast application on the soil surface. If the fertilizer contains only N, subsequent rains will move the fertilizer into the soil, and broadcast application may be as effective as any other placement. Urea is rapidly hydrolyzed to NH_4^+ under warm and moist conditions. When urea is broadcast under alkaline conditions, however, the NH_4^+ reacts with OH^- to form NH_3 and water with great potential for loss of N by volatilization. It is not uncommon to lose over 30% of the N from surface application of urea. As a consequence, the effectiveness of a broadcast application will be increased if the urea is leached into the soil or incorporated by subsequent tillage.

It has been found that up to 30% of N from urea applied to the surface of turf can be lost even if irrigated immediately after application.

Phosphorus is readily fixed in most soils and thus moves very little from the point of application. Broadcast-applied P is therefore less effective because few roots are active in the immediate surface soil, especially when the surface soils dries. In many forage fields, however, the surface of the soil may remain moist under dense foliage for a considerable part of the growing season, and the uptake of P is effective. For insoluble phosphates, in the form of rock phosphate, broadcast application, and subsequent incorporation into the soil, is more effective than band placement because it maximizes soil contact and the opportunity for dissolution. Gypsum and elemental S are common carriers for S and are commonly broadcast because SO_4^{2-} is mobile and readily moves into the soil with water.

All airplane and helicopter applications are by broadcasting. In the U.S., aerial seeding and fertilization of flooded rice fields is routine. Rough terrain and inaccessibility of many forests makes aerial application popular. In these instances, urea is commonly applied, and a large prill size that drifts little is used and can be uniformly distributed.

14.1.4 Combining Broadcast Application with Tillage Operations

Fertilizer use in the past has greatly increased soil test levels for P, and the overall fertility level has increased for many soils. Thus, to some extent, the negative effects of P fixation from broadcast applications have been diminished.

The lower cost of fertilizer relative to labor and other application costs has encouraged higher rates of fertilizer use as well as less attention to effective placement and proper time of application. In the fall and winter, more labor is available, and perhaps soils are firmer for application with bulk spreaders. The price of fertilizer is also more attractive in the off-season. The broadcasting of bulk fertilizer and plowing under the fertilizer is therefore popular. Plowing and subsequent tillage operations mix the fertilizer throughout the plow layer and contribute to effective use of the nutrients.

14.1.5 Application in No-Till Systems

In humid regions, reduced tillage to lessen soil erosion and costs of production has proved popular for row crops, especially for corn in the southern Corn Belt. In no-till planting systems, there is no conventional plowing. Crop residues and organic matter accumulate at the soil surface. Weed and competing vegetation are controlled by chemical herbicides. Only a 2.5- to 7.5-cm wide strip of soil is disturbed for each planted row. Fertilizer applications are mainly by broadcast on the surface. The system builds up a concentration of organic crop residues and carry-over nutrients in the upper few centimeters of soil. There is greater microbial and root activity in this upper soil zone and greater opportunity for immobilization and mineralization of N. The possibility exists for immobilization of N to decompose crop residues in the upper soil layer. The acidic effects of both acid-forming N fertilizers and of organic matter mineralization are, similarly, concentrated in the few centimeters of surface soil. These effects would appear to have an influence on crop yields; however, yields from no-till have generally not been significantly different from yields of conventional tillage systems. Perhaps the sameness of yield is related to other affects of no-till planting.

In no-till planting there is more water available for crops because infiltration is greater and surface evaporation less. The accumulated organic matter at the surface acts as a mulch and causes lower soil temperatures in the spring. The effect of lower soil temperature, especially in the northern states, has been to stimulate a response to some starter K fertilizer when soil K tests were less than optimum. In addition, the compaction of no-till soil and its effect of reducing aeration has apparently reduced K availability in some experiments. These findings confirm the complexity of the soil-plant-climate system; perhaps additional research will give a clearer picture of the effects of no-till on soil properties and response of crops to fertilizers. In the future, the development of better planters with which to place fertilizer in the soil, together with an occasional plowing, will likely overcome any negative aspects of nutrient and pH stratification.

14.1.6 Fertigation

The application of fertilizers in irrigation water is called *fertigation*. Anhydrous ammonia was first directly applied as a fertilizer in the 1930s in California when anhydrous ammonia was added to irrigation water. The two major problems of fertigation have been the inability to distribute nutrients uniformly, because water distribution is not uniform, and fixation of P in the surface soil where root activity is low. There has been much progress in solving both of these problems so that, today, applying fertilizers in the irrigation water is often the most convenient and inexpensive method of application. Fertigation may also offer more flexibility in timing and in rates of application and better control of leaching so that fewer mobile nutrients are lost from sandy soils. Disadvantages include loss of nutrients in runoff water; volatilization loss of N from sprinkler-applied NH_3 and aqua ammonia, especially on calcareous soils; the precipitation of incompatible fertilizer materials; and the clogging of nozzles and emitters.

The major management decisions for fertigation relate to the nature of the fertilizer materials and method of water distribution. Water distribution is by open or gravity systems, such as furrow or flood, or by closed systems which include sprinklers and drip or trickle systems. Aqua ammonia and NH_3 are subject to serious volatilization loss from both water and soil surface in both closed and open systems. Losses from sprinkler application in excess of 50% have been reported. The loss is enhanced by alkalinity of water and soil, by air turbulence, and by increases in temperature, rate of application, and length of exposure. Ammonia added to irrigation water increases the pH of the water and, in the presence of Ca^{2+}, Mg^{2+} and HCO_3^-, causes Ca and Mg salts to precipitate, which can clog closed-system equipment. Urea-ammonium nitrate (UAN) solutions are popular because the N is less volatile and there is more efficient use of N. If dry N carriers are dissolved in water and used in closed systems, they must be free of any coating materials which will plug or clog nozzles and emitters.

Little P is applied by fertigation because the P tends to accumulate at the soil surface where root activity is minimal. In alkaline water, the P may precipitate with Ca and Mg, again causing a clogging problem. Acids can be added to water to retard precipitation of salts; however, the acids are corrosive and require special handling. Research is being conducted to evaluate the effectiveness of acids for reducing ammonia loss and the precipitation of salts in the application equipment. The K carriers, KCl and K_2SO_4, can be added to irrigation water with no apparent precipitation problems. Soluble sulfate carriers are used to add S. Some soluble chelated forms of micronutrients are also used.

A major limitation of fertigation is the nonuniform distribution of water and, therefore, nonuniform distribution of nutrients in open systems. The major advantage of fertigation is the increased opportunity to add fertilizer. Since N is the major nutrient applied, this means better timing of N in relation to plant needs during the growing season. In addition, on sandy soils with pivot sprinkler systems, there is greater control over nitrate N loss by leaching and over groundwater contamination.

14.1.7 Foliar Application

Foliar application can be used for the quick correction of a nutrient deficiency in plants. The amount of fertilizer nutrients that can be applied to the foliage is limited by the solubility of the materials, salt tolerance of the foliage, and the amount of liquid that adheres to the leaves. Most of the interest in foliar sprays is for the application of micronutrients. Some of these, Mn and Fe for example, are rapidly fixed or converted into insoluble form in alkaline soils, and plants have a very small requirement. Often, there must be repeated sprayings, which limit the use to the more valuable crops, such as tree fruits. If the crop is irrigated with overhead sprinklers, the micronutrient can be applied in the irrigation water. Pineapples in Hawaii are regularly sprayed with ferrous sulfate to supply iron. Urea is readily absorbed by leaves and pineapples have high tolerance for foliar salts. As much as 75% of the total N need of pineapples have been supplied from foliar application of urea. The high plant requirements for N, P, and K generally make foliar application of limited use for the application of fertilizers containing these nutrients.

During the seed-filling stage of grains, there is considerable movement of N, P, and K to the seeds. These nutrients are quite mobile in plants; however, leaves are depleted of these nutrients. Some nutrients like Ca, Mg, and Mn are not effectively translocated from leaves to seeds. Thus, it would seem that the addition of certain nutrients by foliar application during seed-filling time might have a significant effect on grain yield. Some research on the application of foliar nutrients during seed-filling time of grains has sometimes shown a significant increase in yield. Some general conclusions from this work are as follows:

1. Urea is the most effective form of N for foliar application. The urea should not be applied when the sun is shining brightly and biuret in the urea may cause leaf burn and reduce yield. The maximum rate of urea that can be applied at one time without causing serious leaf burn is N at a rate of 22 kg/ha (20 pounds/acre).
2. Where foliar fertilization has increased grain yields, the four nutrients, N, P, K, and S, were applied. A ratio of 10N-1P-3K-0.5S was the most effective for soybeans.
3. Including micronutrients or the less phloem-mobile nutrients (Zn, Cu, Mn, B, Ca, Mg) did not increase soybean yields.

14.1.8 Deep Soil Placement

If surface soils are dry and subsoils are moist, the greater water potential in the subsoil encourages greater root growth there and subsequently greater use of the nutrients placed in the subsoil. Root penetration is inhibited into soil layers devoid of Ca. In some very acid and highly weathered soils, deep placement of Ca is necessary in order to promote deep rooting and the use of subsoil water. Deep placement as used here generally means placement in the subsoil; it is expensive and only used occasionally (Figure 14.2). It is to be expected that a plant's use of nutrients placed deep is less efficient because maximum root activity is usually in the plow layer.

Figure 14.2 Fertilizer applicator mounted on a disk plow for placing lime and fertilizer up to 75 cm deep in the soil.

14.2 FERTILIZER RECOMMENDATION PHILOSOPHIES

Fertilizer recommendations based on soil tests are of three types: (1) now that many soils test high for P and K, a maintenance philosophy is employed; (2) soil tests are used as a measure of nutrient sufficiency and are the basis for determining the amount of fertilizer needed; (3) fertilizer recommendations are based on a consideration of cation balance.

14.2.1 Maintenance Fertilizer Recommendations

One of the most striking effects of human occupancy and the agricultural use of land is the soil's increased content of phosphorus. Ancient campsites and kitchen middens have been located in the U.S. and Europe by a systematic sampling of soils and a determination of their P content. Thick, dark-colored A horizons with properties similar to mollic horizons, except that they contain 250 ppm or more of P_2O_5 soluble in 1% citric acid, qualify as *anthropic* horizons. Anthropic horizons are formations greatly affected by human activity. In villages and campsites, P is discarded as bones and other refuse, and the P is subsequently fixed and accumulates in the soil. Agricultural land accumulates P from long-time manuring and fertilization. In Michigan, the median soil test for P increased from 23 to 105 pounds/acre between 1962 and 1986. Large increases in soil P over a relatively short time for a variety of major crops are given in Table 14.1.

Table 14.1 Median Phosphorus Soil Test Levels for Selected Counties and Major Crops in Michigan

County	Major crop	P in soil, pp2m			
		1972	1976–77	1979–80	1985–86
Grand Traverse	Fruit	48	58	83	86
Montcalm	Potatoes	137	178	246	242
St. Joseph	Corn	82	109	171	164
Ingham	Cash crop	14	62	84	97
Ontonagon	Forage	17	18	34	95

Sources: From Ellis and Olson, 1986, and J. Dahl, personal communication.

The extent of P carry-over depends on soil fixation and crop removal. In Illinois, the addition of 9 pounds of P_2O_5 (4 pounds of P) increases soil tests by about one pound/acre. Thus, for many soil and cropping situations the P soil test may become high in a few years, for a high test is frequently only 30 to 60 pounds/acre. It is very common for gardens to test very high in P. Some state soil-testing laboratories now report that 50% or more of the soil samples test high or very high. Potassium carryover is also very common and related to soil fixation and crop removal. Several states report that about 4 pounds of fertilizer K_2O (3.3 pounds of K) will increase the K soil test 1 pound/acre. But the increase in soil-test K has been much less dramatic than the increase in soil-test P. Nitrogen carry-over is strongly related to leaching and denitrification losses. There may be significant carry-over of nitrate N from one year to the next in soils with aridic and ustic soil moisture regimes. Soils in general, however, do not show continuing increases of available N to very high levels as they do for P and K. Thus, soils will rarely produce maximum or high yields of grain without additional N fertilization.

Because soil tests are frequently high for P and K, it is becoming popular to make maintenance recommendations for these nutrients. The amount recommended is based on an estimation of about 1.5 times the amount removed in the harvested crops. Soil tests every two to five years are recommended to determine whether recommendations need to be adjusted.

14.2.2 Nutrient Sufficiency Recommendations

Sufficiency recommendations are based on the soil-test correlation and calibration procedures discussed in Chapter 13. The soil-test value is used to predict or estimate the amount of fertilizer needed to achieve the desired yield goal.

14.2.3 Basic Cation Saturation Ratio

The basic cation saturation ratio (BCSR) or cation balance method attempts to adjust the distribution of the exchangeable Ca, Mg, and K to about 65, 10, and 5% saturation, respectively. For high activity (HAC) soils, this results in 80% saturation of the cation exchange capacity (CEC) determined at 7.0 or above. The percentage saturation ratio of Ca:Mg, Ca:K, and Mg:K would be 6.5, 13, and 2, respectively. The ratios for the plow layers of fields on which world-record corn yields were produced in 1973 and 1977 were 4, 23, and 6 in 1973 and 3, 18, and 5 in 1977. Research on soils with a significant amount of CEC and in which the amounts of XCa and XMg are large relative to annual crop needs (Alfisols and Mollisols) has failed to show a significant relationship between crop yields and percent saturation of Ca, Mg, and K. Very wide differences in the ratios of these exchangeable cations with one another have been observed with no negative effect on yields. Recommendations based on these ratios tend to result in overfertilization and increase the cost of production.

In some situations, the ratios are important. The excessively high XMg relative to the other exchangeable cations is associated with infertility in serpentine soils. High K fertilization can bring on a Mg deficiency for some crops growing on soils marginal in content of XMg. Oxisols have very little CEC and small absolute amounts of XCa..Na. Crops growing on Oxisols are more sensitive to small soil-fertility differences as compared to crops growing on Alfisols and Mollisols. One example of a recommendation for cropping on Oxisols is an absolute minimum amount of K equal to 0.1 cmol/kg of soil and a relative minimum amount of K equal to 2 to 3% of the XCa..Na.

14.2.4 Summary Statement

An eight-year study in Nebraska was conducted on four different soils and locations for corn production. Soil samples from each location were routinely sent to five different soil-testing laboratories for fertilizer recommendations. All three philosophies were represented, and all of the corn yields produced were similar. There were, however, large differences in kinds and amounts of fertilizer recommended. Consequently, there were large differences in the fertilizer costs. The results generally supported the nutrient sufficiency philosophy as being agronomically and economically sound. Similar conclusions have been developed in Kentucky which show that there is considerable differences in cost but little difference in yields.

14.3 FERTILIZER AND PLANT-WATER RELATIONS

Evapotranspiration (ET) is a physical process controlled by the meteorological conditions or the amount of heat available for evaporating water from the soil and vegetation. When water is nonlimiting, the ET or water consumed by a crop is a function of meteorological conditions and not rate of plant growth. Under these conditions, any management practice that increases yield will increase efficiency of water. When water is limiting and soils become dry, soil nutrient availability and water use is less efficient.

14.3.1   Water-Use Efficiency

The water consumed for each unit of yield is a measure of efficiency with which water is used. In comparisons of fertilized and nonfertilized fields or experimental plots, the heat available for evaporation of water can be considered a constant. As a consequence, the ET would be expected to be the same on both control and fertilized plots. One agronomic modifier of the actual ET is the extent of vegetative cover. Once an adequate vegetative cover is established, however, additional plant growth or cover has no or very little effect on ET. Larger plants can intercept more advective or horizontally transferred heat, but this effect is small. The result is that any yield increase brought about by using fertilizer invariably results in increased efficiency of water use. Reports of 50% or more increase in crop yield per inch of water used are common where the response to fertilizer is large. Increased water-use efficiency is also verified by the causal observation that doubling the average yields of a crop in a state does not double the use of water or double the water deficit. As water for irrigation becomes more expensive and less available, it becomes more important to maximize production from the limited supply of available water.

It has been shown that correction of Mn deficiency can result in a decrease in total water use as well as an increase in yield. For some crops, cauliflower for example, this effect was associated with an increased thickness of wax layer on the leaves with Mn fertilization.

14.3.2   Effect of Water on Nutrient Availability

Plants can obtain their nutrients and water from a very small soil volume if both nutrients and water are highly available. When soils dry, however, the matric water potential and hydraulic conductivity decrease. Decreased hydraulic conductivity reduces the movement of water through soil to roots, thereby reducing the rate of water uptake and rate of movement of nutrients to roots by mass flow and diffusion. Plants under water and nutrient stress photosynthesize less. Both top and root growth may be restricted. Droughty conditions, therefore, may result in restricted root growth and decrease the potential supply of nutrients (and water) positionally available for uptake. Increased soil fertility, on the other hand, may increase the growth rate and the extent of the root system, thereby increasing the amount of nutrients and water that are positionally available for uptake (Figure 14.3).

The Cisne soil referred to in Figure 14.3 is a claypan soil, Albaqualf, with smectitic clays in the claypan. Research in Missouri on similar soils showed that breaking up the claypan with deep tillage during the dry season produced only temporary results. Fertilization, however, proved to be the most effective means of increasing deep root growth and yields.

As soils dry, there is a decrease in the movement of nutrients to the roots by mass flow and diffusion. In dry soils, compared to moist soils, fewer nutrient ions move to roots by mass flow because there is less water moving. The diffusion path for nutrients (and water) becomes more tortuous as the soil dries, so that

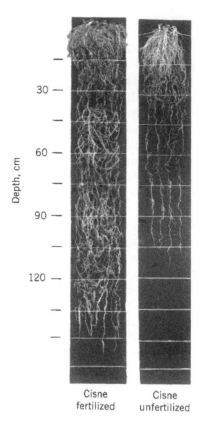

Depth, cm

30 —

60 —

90 —

120 —

Cisne
fertilized

Cisne
unfertilized

Figure 14.3 Fertilizer greatly increased deep root growth of corn in the claypan of the Cisne silt loam (Albaqualf). (Photograph courtesy J.V. Fehrenbacher.)

fewer ions move to roots by diffusion. And the amount of nutrients released when organic matter mineralizes decreases when soils dry, affecting the supply of N in particular and sometimes decreasing B to the point of deficiency. There is an interaction between soil fertility and soil water so that it may be difficult to determine in many times of drought whether yields are reduced more by a lack of water or by a lack of nutrients.

14.4 ECONOMICS OF FERTILIZER USE

Two major decisions relevant to the economics of fertilizer use are the price of fertilizer relative to the price of crops and the price of fertilizer relative to the price of other factors of production. In recent decades, the price of labor, machinery, and land have increased much more rapidly than the cost of fertilizers. As a consequence, farmers have tended to substitute fertilizer for other factors of production, and this has accounted in large part for the rapid increase in fertilizer use since World War II. In the past few years, however, the effect of increasing soil-tests values and the adverse economic situation have moderated the trend.

14.4.1   Profit Maximization from Fertilizer Use

The entire plant-growth curve is sigmoidal. In the range where decisions about fertilizer use are made, the addition of fertilizer causes increasing growth at a decreasing rate. Returns are diminishing. The first units of fertilizer produce greater yield response than later units. Profits from the use of fertilizer are at a maximum when the last or marginal unit of fertilizer produces a yield increase whose price is equal to the cost of the last unit of fertilizer. The ratio of the price of the crop to that of the fertilizer is 1.0, and the point of maximum profit occurs at a rate less than that needed to produce the maximum yield, as shown in Figure 14.4.

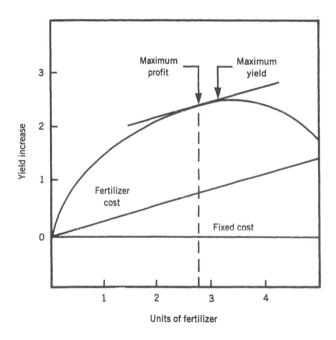

Figure 14.4 The maximum profit from using a fertilizer occurs at a yield that is less than the maximum yield.

There are several other factors that affect the actual decision regarding fertilizer rate. Since the marginal return from additional units of fertilizer becomes less and less profitable, having an alternative, more profitable investment in which to put their money would influence farmers to apply fertilizer at less than the point of maximum profit. Sometimes a lack of capital and risk are important factors. In many developing countries, there is little or no opportunity to make profit from the use of fertilizer because the cost of the fertilizer is too high relative to the price of the crop.

14.5 ENVIRONMENTAL CONCERNS

There are two important environmental concerns regarding fertilizer use. The first is the effect of N and P on the eutrophication of surface waters and, second, the accumulation of nitrate in both surface and groundwater.

14.5.1 Nitrogen

The rate and time of application are the two major issues in the use of N fertilizers. Generally, the rates recommended by soil-testing laboratories are acceptable if the soil profile is low in nitrate. In soils with ustic and aridic soil moisture regimes, the nitrate levels in the soil should be measured. The quantity in the profile should be subtracted from the normal recommendation.

Timeliness of application is an important concern in humid regions. To be environmentally acceptable, N should be applied so it will be retained in the root zone during the growing season. Potential for nitrate leaching is created when more NO_3^- exists in the soil than plants can use. The extra NO_3^- is leachable as shown in Figure 14.5. Thus, if N fertilizer rates are just sufficient to meet yield goals, little potential for nitrate leaching is created. Many farmers, however, set optimistically high yield goals and tend to use more N fertilizer than is needed.

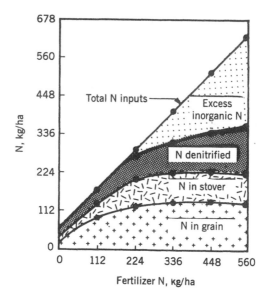

Figure 14.5 Extensive losses of nitrogen by denitrification and loss of excess inorganic nitrogen by leaching occur from rates of N application in excess of plant needs. (From Boswell, Meisinger, and Case, 1985. Used by permission of the Soil Science Society of America.)

Excessive nitrate in drinking water can produce various disorders in livestock and methemoglobinemia, lack of oxygen in blood, in infants. Methemoglobin is a blood pigment that differs from hemoglobin. The U.S. public health standard is 10 ppm (10 mg/L) N as nitrate in water.

In the early 1970s, it was discovered that N_2O released into the atmosphere causes reactions that decrease the amount of ozone (O_3). Loss of ozone reduces the atmosphere's ability to screen out ultraviolet radiation. The production of N_2O is a natural event that occurs during denitrification. Anything that increases N fixation, including N fixed by legumes, increases the potential for more N_2O production. Managing soil pH al levels 6.5 or above decreases the quantity of N_2O produced during denitrification. At the present time, production of N_2O appears to be a valid concern, but the magnitude of the problem is not known. Any fertilizer-use practice, however, that contributes to more efficient use of N will minimize N_2O problems.

14.5.2 The Pre-sidedress Nitrate Nitrogen Test

The potential for nitrate leaching is greatest when soils are mineralizing N rapidly during the growing season. Heavily manured soils are examples of situations where it is difficult to estimate the precise amount of nitrogen fertilizer to apply at or before planting time. The pre-sidedress nitrate test discussed in Chapter 6 measures the amount of nitrate in the soil just prior to the time that it would be sidedressed with N. Therefore, the test measures nitrate that has been mineralized in addition to preplant or planting-time fertilizer. Adjusting the N fertilizer rate at this stage of growth helps to assure maximum yields with a minimum risk of leaving excess nitrate in the soil profile at the end of the growing season.

14.5.3 Phosphorus

Generally, the P concentration of the soil solution is very low and P remains quite immobile in the soil. In recent years, the carry-over of P, however, has greatly increased soil tests. The relationship between the P soil test and the concentration of P in solution for two soils is shown in Figure 14.6. The Hoytville is fine textured and the Hillsdale is sandy soil. A given amount of P in the sandy soil made the concentration of P in solution much higher. Thus, high soil tests on some soils cause concern for eutrophication from P in runoff. Recent research has shown significant downward movement of P in sandy soils with very high soil-test levels. The downward movement is not an environmental threat unless it increases the concentration of P in tile drainage or increases the P concentration in shallow groundwater.

The practice of no-till cropping has introduced an additional way in which P can be lost to the environment. Continuous no-till cropping builds up greater amounts of P in the immediate soil surface and the potential for greater movement of P by erosion. On the other hand, runoff and erosion are generally much less than where conventional tillage practices are used because more crop residues remain on the soil surface.

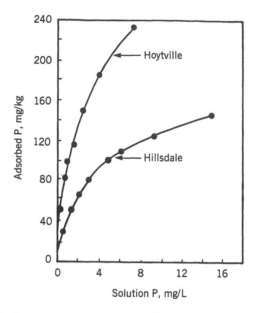

Figure 14.6 Partitioning of phosphorus between the solution and the solid phase. (From Ellis and Olson, 1986.)

14.6 SITE-SPECIFIC FERTILIZER APPLICATION

Methods have been developed to allow for computer-controlled fertilizer application on a field basis. A field can be mapped either by soil type or by soil test and the map used to alter the rate of fertilization based on the need of each specific site in the field. This offers the possibility of much more precise fertilizer application and greater fertilizer-use efficiency. Such practices should reduce the potential for environmental degradation due to overfertilization.

REFERENCES

Aldrich, S. R. 1984. Nitrogen Management to Minimize Adverse Effects on the Environment. *Nitrogen in Crop Production.* Am. Soc. Agron., Madison, WI.

Barber, S. A. 1984. *Soil Nutrient Bioavailability.* John Wiley & Sons, New York.

Boswell, F. C., J. J. Meisinger and N. L. Case. 1985. Production, Marketing and Use of Nitrogen Fertilizers. *Fertilizer Technology and Use.* 3rd ed. Soil Sci. Soc. Am., Madison, WI.

Ellis, B. G. and R. A. Olson. 1986. Economic, Agronomic and Environmental Implications of Fertilizer Recommendations. *North Central Regional Res. Pub. 310.* Mich. Agr. Exp. Sta., East Lansing, MI.

Follett, R. H., L. S. Murphy and R. L. Donahue. *Fertilizers and Soil Ammendments.* Prentice-Hall, Englewood Cliffs, NJ.

Foth, H. D. 1974. *Effect of Plow Depth and Fertilizer Rate on Yields of Corn, Barley, Soybeans and Alfalfa and on Soil Tests.* Res. Report 241. Mich. Agr. Exp. Sta., East Lansing, MI.

Gilliam, J. W., T. J. Logan and F. E. Broadbent. 1985. Fertilizer Use in Relation to the Environment. In *Fertilizer Technology and Use.* 3rd ed. Soil Sci. Soc. Am., Madison, WI.

Iowa State Univ. Coop. Ext. Ser. 1982. *General Guide For Fertilizer Recommendations in Iowa.* Agron. Dept., Ames, IA.

Nat. Res. Council. 1972. *Soils of the Humid Tropics.* Nat. Acad. Sci., Washington, D.C.

Ohio Coop. Ext. Ser. 1986. *Ohio Agronomy Guide.* Bull. 472. Agronomy Dept., Columbus, OH.

Olson, R. A., K. D. Frank, P. H. Grabouski and G. W. Rehm. 1982. Economic and Agronomic Impacts of Varied Philosophies of Soil Testing. *Agron. J.* 74:492–499.

Plank, C. O. 1985. *Soil Test Handbook for Georgia.* Ga. Coop. Ext. Ser., Athens, GA.

Randall, G. W., K. L. Wells and J. J. Hanway. 1985. Modern Techniques of Fertilizer Application. *Fertilizer Technology and Use.* 3rd ed. Soil Sci. Soc. Am., Madison, WI.

Robertson, L. S., D. D. Warncke and D. L. Mokma. 1978. Test Levels in Profiles of Two Soils Producing World-Record Corn Yields. *Research Report 363.* Mich. State Agr. Exp. Sta., East Lansing, MI.

Terman, G. L. 1982. Fertilizer Sources and Composition. *Handbook of Soils and Climate in Agriculture.* CRC Press, Boca Raton, FL.

Tisdale, S. L., W. L. Nelson and J. D. Beaton. 1985. *Soil Fertility and Fertilizers.* 4th ed. Macmillian, New York.

Univ. Illinois Coop. Ext. Ser. 1984. *Illinois Agronomy Handbook 1985-1986. Cir. 1233.* Agron. Dept., Urbana, IL.

Viets, F. G. 1961. Fertilizers and the Efficient Use of Fertilizers. *Advances in Agronomy, Vol. 14 .* Am. Soc. Agron., Madison, WI.

Viets, F. G. 1971. Fertilizer in Relation to Surface and Ground Water Pollution. *Fertilizer Technology and Use.* Soil Sci. Soc. Am., Madison, WI.

Wolcott, A. R., H. D. Foth, J. F. Davis and J. C. Shickluna. 1965. Nitrogen Carriers: Soil Effects. *Soil Sci. Soc. Am. Proc.* 29:405–410.

Woodruff, C. M. and D. D. Smith. 1946. Shattering and Subsoil Liming for Crop Production on Claypan Soils. *Soil Sci. Soc. Am. Proc.* 11:539–542.

Index